D0536601

ALTERNATIVE FARM ENTERPRISES

ALTERNATIVE FARM ENTERPRISES

A GUIDE TO ALTERNATIVE SOURCES OF INCOME
FOR THE FARMER

Bill Slee

FARMING PRESS

First published 1987
Second edition 1989

British Library Cataloguing in Publication Data

Slee, Bill
    Alternative farm enterprises: a guide to
    alternative sources of income for the
    farmer.
    I. Farm management
    I. Title
    630'.68   S561

    ISBN 0–85236–192–0

Cover design by Hannah Berridge

Cover photograph taken near the
White Horse of Kilburn, North Yorkshire by John Edenbrow

Phototypeset by Input Typesetting Ltd, London
Printed in Great Britain by Biddles Ltd, Guildford & Kings Lynn

CONTENTS

LIST OF FIGURES

LIST OF TABLES

FOREWORD

The first edition of *Alternative Farm Enterprises* was published as the debate about agricultural restructuring and farm diversification was heating up in the United Kingdom. Two years later the principal assertion – that a more market oriented industry would need to look at a more diverse set of enterprises in the future – remains valid.

However, the context of the debate has changed. Like problems of ozone depletion and acid rain, farm diversification is an international issue. The New Zealand sheep farmer, the Minnesotan grain producer and the British dairy farmer all face a common problem: how to respond to policy instruments which are seeking a better balance between supply and demand. Diversification is one of the options.

The second way in which the context of the debate has changed is that in some countries, the UK in particular, policy instruments are now in place that will affect, directly and indirectly, the extent to which alternative enterprises are taken up. There is a need to review the effectiveness of these policies and to assess the extent to which their operation is likely to impact upon those who have already travelled on the road to diversification.

The form which agricultural restructuring takes will vary greatly from place to place, from farmer to farmer. Cultural differences will colour not only the policy instruments but also the individual responses. But the imperatives that are everywhere directing the adjustment of the agricultural sector in developed Western style economies are essentially the same.

The individual farmer is likely to be perplexed by a policy environment which has become more volatile. What can he do? He can contribute to the defence of his way of life through his union or representative organisations. He may pen letters to the farming press condemning the activities of environmentalists, health experts or politicians. However, the crucial response will be not what farmers say or write but what they do. The challenge to the farmer is how to guide his business through what is likely to be a difficult period of adjustment in the next few years. Over-optimistic slogans relating to expansion must be replaced by realistic slogans relating to survival.

It is relatively easy to identify appropriate business strategies for farmers in times of policy stability. However, as policies affecting farmers enter a period of rapid change, the scope for individuals making ineffective adjustments to their businesses increases. It may not be sufficient to resort to the traditional agricultural response of belt tightening. What is needed is a recognition of changes in market signals and government policy which create a need for a more fundamental review. The review process at a collective or individual level must consider not just conventional agricultural enterprises but also the scope for establishing alternative enterprises on farmland. Options for farmers that were irrelevant in a time of high levels of price support for conventional products may be lifelines in a restructured policy framework.

The idea of an alternative enterprise must be clarified at the outset. Conventional farming is assumed to embrace the primary production of food and fibre with minimal processing on the farm. To define alternatives in such a way as to include only those enterprises that fall outside *all* primary production of food and fibre would preclude the consideration of many enterprises that are considered as 'alternatives' by the farming community, its advisers and the public at large. The generally accepted conception of what is an alternative farm enterprise is one which is not covered by price support arrangements (where they exist for other agricultural products) and which embraces a vast array of possibilities. These range from farming alligators in Florida to farm zoos and parks the world over. Some of these alternatives will cease being regarded as alternatives with the passage of time and become mainstream enterprises. There will be a process of accretion and erosion to the list of alternatives with the rate of change being influenced by the policy climate and the technical and entrepreneurial ingenuity of the industry.

It is important to recognise that diversification does not necessarily take place on the farm. Ruth Gasson has shown in her book *Gainful Occupations of Farm Families* that these off-farm sources of income represent a form of business diversification that embraces more farmers than on-farm diversification. Off-farm diversification may be especially relevant for farm households where opportunities for on-farm diversification are extremely limited. Equally, though, the off-farm employment evidenced in research surveys may represent the full-time business activities of hobby farmers. The range of off-farm sources of income is as varied as the opportunities in the economy as a whole, although there are likely to be geographical variations in the extent of these opportunities.

Neither on-farm alternatives nor off-farm sources of income are especially new. All, however, create a need to look at business activity in what is to many farmers a radically different way. Farmers must look to consumers' needs and respond accordingly to get rid of the production orientation that has shackled the industry. The search for alternative enterprises must be guided by a new philosophy.

Whilst politicians conduct protracted wranglings over international trade in agricultural products and Eurocrats and their cosy coterie of researchers develop their own perceptions of what they term *pluriactivité* (multiple job holding), farm families and upstream and downstream industries are being forced to reappraise their situations.

It is doubtful whether the intensive arable farmer from East Anglia or Iowa, the dairy farmer in southwest England or Ireland or the stock rearing farmer in New Zealand or the north of Scotland can ignore the changing climate of policy. In order to obtain the incomes they want at an acceptable level of public expenditure to society as a whole, adjustment will be necessary. All must think hard about the range of goods and services they produce for an increasingly discerning public. This book is an attempt to guide the farmer, landowner or student of rural affairs through the factors that have caused the adjustment need and the way of thinking that should influence the response on the ground.

ACKNOWLEDGEMENTS

I would like to thank the following colleagues who have provided both comments and ideas: Brian Camm, John Kirk, Richard Soffe and Martyn Warren of Seale Hayne College; Watson Bell of Agricultural Management Consultants. Matt Merrick, Charlie Wooding and David Perryman are thanked for photographs. Kev Theaker is thanked for his help with library searches.

I am particularly indebted to those who run the alternative enterprises which I have visited in Scotland, the Welsh Borders, the Midlands, southwest England and overseas, from whom I have learned so much.

My greatest debt is to my family, particularly my wife, who typed the manuscript and commented on the text.

DR BILL SLEE
Bishopsteignton
September 1989

ALTERNATIVE FARM ENTERPRISES

Chapter 1

AN INTERNATIONAL PERSPECTIVE ON ALTERNATIVES

For the farmer on the ground in the United Kingdom it may be difficult and seem unnecessary to make connections between the problems faced by farmers in North America, Australasia or even elsewhere in the European Community (EC). Different countries have different cultures, different farm structures and different policy instruments with respect to their farm sectors. Yet there is an important connection. Farmers in all these places are struggling to adapt to a changing policy context which is the result of an international process. Whilst the responses may be mediated by national and local factors, the overriding cause is an international one, namely the apparently inexorable upward trends in the output of the farm sector in developed Western economies. There have been hiccoughs in the upward trend caused by a run of poor harvests in Europe in the mid 1980s and the exceptional drought in 1988 in parts of the United States. But the long-term trend has been one of increasing costs of supporting agriculture because of the increased capacity to produce. In addition, there have been other wide-ranging criticisms of the agricultural sector, often nationally idiosyncratic, which are leading to the development of new and often radically different policies for the agricultural sector and the countryside as a whole.

The UK has been a major contributor to the debate on policy reform and farm diversification. The contributions of UK farmers and policy makers to the development of alternative enterprises provide most of the examples explored in this book. It would, though, be blinkered not to take a wider view. New Zealand farmers faced the chilling winds of a market oriented agricultural policy in

1

the early 1980s, precipitated in part at least by the global restructuring of agricultural product markets. Their experience may yield useful lessons. More recently, farmers in various parts of the US have begun to inquire into, and develop, alternative enterprises as part of their adjustment process. Their responses merit consideration, for part of their policy package in the past has been 'set-aside', which has just been tentatively introduced to Europe. The set-aside schemes clearly impinge on and interact with diversification.

The agricultural sector has attracted greater public debate and political interest in recent years than its contribution to the national wealth of many countries suggests it ought. There has been an extensive debate about the adjustment needs of the agricultural sector and the rural economy as a whole. There are many interested parties other than farmers, often holding competing views about desirable directions for agricultural change. It is within these shifting sands of opinion and the resultant policy responses that the scope for alternative enterprises must be explored. Franklin[1] has observed that: 'Rural areas are becoming much more multipurpose and social, environmental and recreational issues are coming to the fore.' The implications of this on the search for alternatives are self-evident.

The principal cause of the vigorous debate about future agricultural policy needs has been identified as the capacity of the industry to increase its output at a faster rate than the expansion of demand for its products. This observation is not new. What is new is the extent to which this process has generated what are seen as unacceptable costs by governments and consumers. The observed tendency for supply expansion to outstrip the growth of demand would happen in the absence of any government intervention, as long as the industry was technologically dynamic and there were limits to the demand for farm products. Governments have aided this process by a variety of measures. Where price support occurs and governments act as guarantors of income in the face of market instability, then the restraint imposed by the market place is thrown to the wind. There are other ways in which government can directly or indirectly influence the expansion of output. A production oriented advisory and extension service has pushed output-increasing technology to the farming community, often backed up by grant aid. The nature of government funded agricultural education and research may fuel the process of supply increases. Finally, governments which have pursued expansionist policies for domestic, political and economic reasons – such as the UK in the late 1970s – may have given a further push to the expansion of supply.

The expansion of output generated few insoluble problems until the 1980s, when many governments, locked into a variety of supply-increasing policies, found themselves facing escalating support costs. These support costs had risen dramatically because of the need to dispose of surpluses into world markets already overloaded with plentiful supplies of food.

The main overflow pipe for output increases for developed agricultural sectors has historically been trade. The UK public's image of developed Western countries being dependent on developing countries for supplies of food has long been a misconception. Developed countries account for about three-quarters of all agricultural exports in the world, the US and EC for almost half. Trade in agricultural products has become the subject of increasingly acrimonious debates, especially between the EC and the rest of the world. Skirmishes, rather than full-blooded trade wars, are the likely outcome but these are indicators that, as a general rule, output increasing policies can no longer be sustained.

There are political, as well as economic, forces at work. Protectionist agricultural policies are increasingly difficult to rationalise when the prevailing political rhetoric is about freeing markets to make them work more effectively. Those countries with particularly high levels of protection, including all EC members and countries like Japan, are likely to find their policies much more difficult to defend. Japan can afford to display oriental imperviousness towards its critics. Its agricultural industry is relatively backward and does not cause any major interference with world trade. The same cannot be said of the EC, which is a major exporter of food but has to subsidise its exports to a massive extent to make them competitive in world markets. In spite of the contradiction between the political rhetoric and contemporary EC policies, no fundamental short-term shift in policy is likely. Political opportunism will ensure this. But the durability of the overriding principles of the Common Agricultural Policy (CAP) and the resulting need to shift surpluses in international markets has generated counter-measures from competitors, as is evident in the export subsidies of the 1985 Food Security Act in the US.

The countries of Australasia have watched the transatlantic squabbling as interested observers. Countries like Australia and New Zealand used to export much greater quantities of food to the UK prior to EC entry. Now they find products from the UK competing with their own in declining world markets. New Zealand farmers in particular have been subjected to a battering by market forces and have had radically to restructure farming practices and reorientate themselves to different products and different markets. Nowhere in the developed Western world is the link between the loss of traditional primary markets and the development of alternative farm enterprises more clearly exposed than in the case of New Zealand.

NEW ZEALAND

The entry of the UK into the EC in 1973 precipitated a period of painful restructuring for New Zealand's farmers. Markets for dairy products were removed or reduced and the effect of the EC's sheepmeat policy has been to undermine the New Zealand producer. However, the problems that have beset farmers in New Zealand in the late 1970s and the 1980s are only partly a result of the UK's accession to the EC. From the late 1970s, domestic economic mismanagement has compounded the difficulties and created the preconditions for the crisis from which the New Zealand farm sector is struggling to emerge in the late 1980s.

New Zealand farming had been dominated by sheep and cattle, reliant on low-cost production to push its products into world markets at highly competitive prices. After the UK had entered Europe, new markets were sought and developed, but the EC still remains important as an export destination. Trade in agricultural products has grown with Japan, the US and Australia. Market outlets for lamb were developed in the Middle East in spite of political instability. This reorientation process and the growing interest in alternative products were much in evidence by the time of the crisis precipitated by the change of government of New Zealand in 1984.

The process of reorientation is evidenced by the development of deer farming. By 1982 there were 185,000 deer held on about 2000 farms, few of which were exclusively deer farms. In the late 1970s, the developing industry had focused on the production of breeding stock and velvet, the latter used in oriental potions and medicines. Breeding stock commanded high prices, which were exaggerated by fiscal incentives. By the mid 1980s, licensed slaughterhouses had been developed and the number of deer coming on stream for venison production is likely to increase dramatically. New Zealand is likely to have a significant share of world venison production in the early 1990s. An added advantage to New Zealand producers is the fact that venison production in the southern hemisphere has been untainted by the radiation fallout from Chernobyl, which has affected the market for venison in northwest Europe. The development of a high-value luxury product from what had been an agricultural pest provides an example of what is likely to prove effective diversification.

The recent emphasis in deer farming is on genetic improvement, seeking to obtain earlier calving, heavier carcases or better antlers. Wapiti and Père David's Deer have both been used in hybridising experiments and may help to retain New Zealand's competitive advantage. However, disease problems may emerge as a result of intensive management and domestication which could yet challenge the development of this nascent industry.

The domestic problems which compounded New Zealand's agricultural adjustment problems began in the late 1970s. A cost–price squeeze was afflicting the farm sector and producers' margins were being eroded by escalating costs in the food processing sector. The generalised response to this situation was for farmers to increase stock numbers to compensate for declining margins. This response would subsequently add to the problems.

Prior to 1978 when the government introduced the Supplementary Minimum Prices Scheme, the producer boards had operated price stabilisation schemes on major agricultural exports. These arrangements subsidised incomes when prices exceeded a specified maximum. This was essentially a means of protecting producers from fluctuations in world prices. In 1978, Minimum Guaranteed Prices were introduced by the government for meat, wool and dairy output. In the early 1980s, prices were set too high, and in the 1981/82 season imposed a cost burden of £200 million on the government.

The new government elected in 1984 was immediately confronted by an unfavourable economic climate with high levels of inflation, an overvalued New Zealand dollar, declining foreign exchange reserves and a large budget deficit. In a radical experiment in free market economics, agricultural support mechanisms were dismantled or phased out, forcing New Zealand farmers to compete in turbulent world markets with minimal support from their domestic government. Incomes slumped, land prices fell and indebtedness rose.[2] Some restructuring of debt has occurred, but roughly 15 per cent of New Zealand's 60,000 farmers faced bankruptcy or severe financial difficulty in the mid 1980s.

This financial nightmare has been the seed bed for some remarkable developments in farm diversification. Horticultural production has expanded. Plantings of kiwi fruit expanded over threefold between 1980 and 1984. New varieties of apples and Asian pears have been planted. Avocados, persimmons and tangeloes have been added to the list of fruit grown. New Zealand's advantage is its ability to provide markets in the northern hemisphere with products out of season. Experimental work is being conducted into truffle production. These products are seen as potential lifelines to an industry that has become all too aware of the weakness of a small primary producer in a world bloated with surpluses.

The animal sector has responded too. The other 'pest' ruminant, the goat, has provided raw material for genetic upgrading experiments to produce high-value animal fibres. Its predilection for weeds makes it doubly valuable as an organic weedkiller. The instability of the existing sources of supply makes fibre production look a useful medium-term prospect.

However, blind optimism is unjustified. The fluctuations in prices for breeding livestock for goat and deer production illustrate the volatility of the free market. Price crashes and crises are inevitable products of market forces. Risk aversion may be a reasonable strategy that will often involve diversification. But until the international power brokers in the northern hemisphere back up their rhetoric about free markets by policies that bring them about, New Zealand farmers will struggle to diversify, to create niche markets and to adapt their industry to what seems like a malevolent and unfair world.

THE UNITED STATES

The farm sector of the US has experienced a profound depression and a major drought in the 1980s. In 1985, the shape of agricultural policy was significantly altered by the Food Security Act. During these upheavals the beginnings of the development of alternative enterprises as an element of policy as well as a farm response can be discerned.

There are parallels between the structures of agricultural policies in the US and in the EC. The federal operation is managed by the US Department of Agriculture whilst the local context is provided for by state specific schemes. At a European level, the CAP provides the framework but there is still room for national schemes and variety in the implementation of EC initiatives.

The 1970s were years of expansion for agriculture in the US under the influence of a federal 'production at any cost' approach.[3] The farm crisis reached its lowest point in 1983 with farmers facing levels of indebtedness unprecedented since the 1930s, and many foreclosures. The crisis precipitated a response in the form of the

Food Security Act of 1985. Some of the measures in the Act were crisis-response measures, e.g. those to relieve indebtedness and to support exports. Others represented a more fundamental stock-taking of the position. The possibility of moving towards the zero option of no market distorting support to the farm sector was being discussed seriously.

'Set-aside' has a long history in the US, dating back to the inter-war period, and, in 1987, 15 per cent of the crop acreage was 'idled'. The 1985 Act retained provisions for set-aside and introduced new measures which were intended to have environmentally beneficial as well as supply-reducing consequences. The Conservation Reserve Program is designed to take highly erodible land out of production. Once land has been classified by the Soil Conservation Service as eligible, farmers are invited to submit bids; the US Department of Agriculture then accepts the lowest bids so as to achieve the lowest cost conservation. There is no guarantee that the land taken out will be the most erodible, but an incentive for farmers to participate is provided by the fact that assistance under commodity support and subsidy programs is linked to participation in the scheme. In addition to the Conservation Reserve Program, there are provisions to protect suitable wetlands.

These provisions offer a type of alternative enterprise where the output is a conservation 'product'. Farmers are thus forced into thinking about the environmental consequences of their actions by being bribed and cajoled into environmentally beneficial practices.

There is widespread recognition that conventional agricultural production will be unable to sustain the present levels of population working in farming. In addition to government policies which have endeavoured to stimulate rural industrialisation by an enterprise zone-type approach, farmers have begun to develop alternative enterprises, sometimes with the help of public agencies. Federal and state initiatives have supported the development of alternatives from federally funded marketing studies of new crops to the development of crop diversification programmes from Oregon to Maine.

The farming press and the general press have also been instrumental in developing awareness of diversification. The most celebrated example of press involvement is the ADAPT 100 conference,[4] staged by Successful Farming in the Midwest in the summer of 1986. Several thousand delegates attended this conference which provided an opportunity to learn about a hundred different ways to diversify, ranging from garbanzo beans to ginseng, from amaranth to angora goats.

The size of the US and the diversity of agricultural enterprises over a vast range of soil and climatic regions, mean that generalisations will inevitably do injustice to individuals, agencies or areas. Two brief case-studies will be used to reflect the diversity of problems and responses.

VIRGINIA

Virginia, on the eastern seaboard, is not an important state in terms of its agriculture. Geographically, it is divided into two principal regions: the Appalachian Mountains and the coastal plain. There are no metropolitan areas, but the northeast corner of the state abuts Washington DC and falls within the shadow of its urban influence.

In the coastal plain, tobacco was a principal crop. The growth of the anti-smoking lobby and the publicity about the adverse health effects of smoking have contributed significantly to the long-term decline of the industry. The principal alternatives in the coastal plain have been field vegetables. There has been a marked expansion in the production of a whole range of field vegetables including snow peas, cantaloupes, broccoli, string beans and peppers. Those who have been promoting these alternatives have also been trying to instil a greater degree of market awareness into growers. The proximity to the large east coast metropolitan markets offers opportunities and there may be market windows (see Chapter 3) when profitable production can be sustained. But it is crucial for the grower to understand the factors that cause price variations and act accordingly.

The development of shii-take (Japanese wood mushroom) production on farms in Appalachian Virginia and on the coastal plain provides a second example of diversification in the state. In the early 1980s, a few entrepreneurs began to produce shii-take commercially, taking advantage of the climatic similarities to Japan and Korea and the abundant supplies of oak cordwood on which the mushroom is grown. A conference in the mid 1980s was attended by 400 growers and interested persons and a growers' association was formed. Major difficulties arose when poor-quality spawn was used by many association members, but the developing industry weathered the crisis.

The state and federal agricultural agencies supported the development of the industry. Multidisciplinary research work with a strong marketing emphasis has been conducted by Virginia Polytechnic Institute for the US Department of Agriculture. Virginia

State University now has a diversification adviser with a particular interest in shii-take production. Local offices of the extension services offer farmers advice on a wide range of alternatives, including shii-take.

The largest shii-take producer in Virginia has produced mushrooms from several hundred thousand logs. This large-scale enterprise operates outside the recently established growers' co-operative which caters for farmers with much smaller enterprises, a number of whom are relatively recent entrants to farming. The Virginian experience illustrates three problems of the development of alternatives. Firstly, there is a need to develop an understanding of the market. Secondly, there is a likelihood that attempts will be made to concentrate production and thus limit the number of potential beneficiaries. Thirdly, recent recruits into farming are likely to have higher levels of market awareness than traditional farmers and thus pick up the windfall profits of early entrants.

These examples of diversification come from a wide array of types of alternatives developed in Virginia. In the hinterland of Washington, 'horseyculture' is much in evidence. Farm shops are found by the roadside and farmers' markets provide opportunities for direct marketing. In the Appalachians, some farmers are growing Christmas trees for the east coast markets. The range of options is huge. The key to the effective development of alternatives depends on the adoption of a marketing approach (see Chapter 3).

MINNESOTA

Minnesota is a characteristic Midwest state with a very important agricultural sector. The farm crisis hit states like Minnesota especially hard and the viability of many family farms is threatened by tightening margins on traditional crops.

The importance of agriculture to the Minnesotan economy has led to a significant public response. The Centre for Alternative Crops and Products is a state initiative based at the University of Minnesota. It aims to provide three principal services to the development of diversification in the state. Firstly, it will act to generate ideas about alternatives and analyse and evaluate the possibilities. Secondly, it will engage in, or assist others in engaging in, research projects relating to diversification. Thirdly, it will disseminate information in a variety of ways to extension personnel and others. It is too early to forecast the effectiveness of this agency or similar agencies that have been set up in other Midwest states.

There are unlikely to be any single products that will meet the demand for alternative enterprises but the success of such agencies will depend on identifying products that can be grown and marketed with competitive advantage in the areas in question. The transference of higher levels of market awareness to growers will be a further indication of the agency's achievement.

The development of the wild rice industry in Minnesota is an example of an alternative enterprise that predates the current high levels of interest in diversification.[5] Wild rice is a native plant that was gathered in lakes and used as a staple food by native Indians. Small quantities of the natural crop were gathered. In the mid 1960s a natural crop failure coincided with plant breeding improvements. Uncle Ben's Inc was instrumental in transforming the industry from the gathering of wild grain to its production as a field crop. Since 1969 the state legislature has funded wild rice research in the University of Minnesota.

The crop requires padi cultivation and 10,000 ha are cultivated. In 1985, there were about sixty-five farmers producing wild rice worth about $32 million in Minnesota. However, since the late 1970s, California has also produced wild rice in the same padi fields used for normal rice production. The wholesale price declined from a peak of $5.15/lb in 1975 to $1.75/lb in 1988. This price decline has occurred at the same time as production increased sevenfold up to 1986 before dropping back slightly in the last 2 years.

Wild rice is a gourmet food, labelled by its promotional agency as 'the caviare of grains'. It sells in a major UK supermarket chain at 5 times its US wholesale value.

The example of wild rice indicates a number of characteristics of alternatives. If the product can be grown readily there is a danger that the alternative will experience the same supply shifts as conventional products. Padi fields are by no means widely available in the US but the development of the industry in California would appear to be associated with declining returns to Minnesotan producers. Secondly, there is a need to promote the product, a task which the Minnesotan Wild Rice Council performs. Finally, the stimulus to the development of the industry and the retail sale price suggests that power in the food chain is not on the farm but with food processors or retailers.

The development of alternative enterprises on farms in the US will continue. The drought year of 1988 produced a catastrophic harvest of conventional crops in many regions. Such events may temporarily divert attention from alternatives, but the potential for increased production and increased budgetary costs remains. The agenda for agricultural restructuring must recognise the need to educate farmers and advisers not just into new crops but into shifting away from a production orientation.

The context in which diversification of farm businesses occurs differs from region to region and country to country. Within the EC, in spite of the unifying structure created by the CAP, the response to diversification has been highly variable, reflecting differences in attitudes as well as in potential. However, the desire of some to invoke market forces contrasts with the political necessity of others to continue supporting the farm sector.

The situation in much of continental Europe is different in that farms are smaller and there are strong traditions of *pluriactivité* or off-farm diversification. This results in strengths and weaknesses. The strength arises from the ability of farmers to operate from a broader and more influential political base. The weakness arises from cost-cutting politicians and Eurocrats contending that agricultural price cutting can be justified where other income sources exist to cushion the blow and cross-subsidise the farm.

The overall solutions offered to the budgetary crisis in the EC often incorporate considerations of alternatives, but they view alternatives at a macro-level and not as an individual response. Their alternatives consist of such enterprises as biofuel production, or biotechnological transformation of surplus products like sugar. But, rather than letting markets rip, the Brussels approach seems bureaucratic. It has been noted that 'all such diversification requires various incentives, aid and technical and economic assistance for the farmers.'[6]

In many ways, the responses to diversification in the UK have more in common with diversification in other English-speaking countries than with Europe. Protectionism in Europe is clearly at odds with contemporary political values in the UK. Farm size is on average much greater in the UK than in most of the EC. This in itself probably leads to a more farm-centred response to alternative enterprises than in the rest of the EC.

Diversification and the development of alternative enterprises should not be viewed in isolation from other changes in the rural economy. The dilution of the traditional rural population by incomers can be found throughout the Western world. People are moving into rural areas not to seek work but an improved lifestyle. This movement can be seen as a threat to the indigenous community. A barricade mentality has developed in some areas. The locals resent the rise in property values (unless they are selling), resent the peculiar demands (unless they can exploit a market opportunity) and are generally aggrieved by the process of counter-urbanisation. The conflict is undeniable but so, too, is the fact that these incomers may be eager consumers for alternative products, as long as their vision of the countryside is not compromised by the production process. The changing nature of the rural population creates a further dimension to the diversification debate that advisers and practitioners ignore at their peril.

REFERENCES

1. FRANKLIN, M. (1988) *Rich Man's Farming: The Crisis In Farming*. Routledge.
2. JONES, J. V. H. (1988) 'New Zealand agriculture: a state of flux', *Farm Management*, **6**, 8, 327–34.
3. LAPPING, M. B. *et al.* (1987) 'Rural Policy and Legislation in the USA 1986', in: GILG, A., *et al.* (eds) *International Yearbook of Rural Planning*. Geobooks.
4. SUCCESSFUL FARMING (1986), *ADAPT 100*, Successful Farming, Kansas City.
5. NELSON, R. N. AND DAHL, R. P. (1986) *The Wild Rice Industry: economic analysis of rapid growth and implications for Minnesota*. Department of Agriculture and Applied Economics, University of Minnesota.
6. HOWE, K. (1987) 'Agriculture in Europe', in: GILG, A., *et al.* (eds) *International Yearbook of Rural Planning*. Geobooks.

Chapter 2

WHY LOOK FOR ALTERNATIVES IN THE UK?

The interest in alternative enterprises as part of the solution to the current problems facing the agricultural industry has grown dramatically in recent years. There are many reasons why those that own and work or have a financial or spiritual interest in farmland should be increasingly interested in farm diversification and restructuring. The current reappraisal of rural policy that is taking place is a response to changing economic, political and social forces in the UK, some of which are reflected elsewhere, and others which have a particularly British flavour.

At various times, and for various reasons, the way in which we use our agricultural land in the UK has been the subject of much debate. Earlier in the present century, two world wars and a major depression have been the primary causes of public debate. In the late 1980s the causes are somewhat different. On one hand, the budgetary cost of the CAP has led member states of the EC to review, and then adjust, the level of support given to farmers. On the other hand, there has been growing concern amongst the public and environmental pressure groups about the effects of modern farming practice on wildlife, landscape and the countryside in general. The combined effect of these two causes is generating a lengthy and, at times, acrimonious review of rural policy in general and agricultural policy in particular.

These two groups of influences on rural policy are likely to remain. Public expenditure constraint is not just a government whim that could disappear at the next election. It is also a European necessity. The rapidly escalating support costs of the late 1970s and the 1980s have forced politicians to act in order to save the EC from bankruptcy. The threat of bankruptcy has been passed to the farmer instead. Coupled with the attempts to restrain increases in

13

support costs, there has been a strong revival of interest in free market economics in the UK as in other parts of the Western world. This has deeply affected government policy in many sectors of the economy. To the free market economist the farming community is an example of an industry that has bloated itself at government's expense.

The second area of concern also shows no signs of diminishing in importance. The widespread concern about the rural environment remains in spite of new legislation in the early 1980s. There have also been marked policy changes in the National Farmers' Union and Country Landowners' Association to attempt to accommodate criticism from an environmental lobby that is growing in strength. A significant and articulate challenge has been made on the firm hold that agricultural representatives have had on rural policy making since 1947. Environmental pressure groups and quangos are a fact of life, and are likely to increase in the future. Often associated with the growth of environmentally green attitudes are changes in attitudes to food and health and to animal welfare.

The existence of alternative enterprises on farms is not new. There are strong historical traditions in some parts of the UK of farmers pursuing other forms of economic activity. Medieval part-time farmers were miners and weavers; early in the present century visitors took 'farmhouse lodgings' around the attractive parts of Dartmoor, and the visiting public was subjected to an early form of direct marketing of such products as butter and eggs. However, the creation of powerful agricultural income support in the years since the Second World War led to an understandable concentration on conventional agricultural products marketed through conventional channels. The present signs of weakening of support for conventional products require farmers to look to alternative enterprises again and build on approaches that have long been suppressed.

There are two sets of related forces to be considered in a review of the prospects for conventional agriculture and alternative enterprises. Firstly, the economic and policy environment affecting conventional agriculture must be examined, for major changes are taking place. Secondly, the changing attitudes of the consuming public must be explored. These interacting forces are creating a new framework in which policies will be designed and implemented, and within which farmers must learn how to survive. Together, these factors explain why alternative enterprises should be considered.

FARM INCOME PRESSURES

Developed industrial or post-industrial nations are not hungry nations. The satisfaction of basic food needs in the UK has ceased to be a widespread problem, although the possibility of the reappearance of nutritional problems amongst certain social groups cannot be ruled out. A growing agricultural industry was a vital contributor to the process of industrialisation, providing food for the industrial population and releasing labour for expanding industries. However, this growing agricultural industry becomes an embarrassing burden with time. As birth rates level out, or even decline, and affluence increases, there is no longer the same demand for additional food. It is easier to satisfy the consumer's food needs than his demands for other products and services, and there is a tendency for farm incomes to come under pressure as economic development proceeds.

The income problems are made worse because the ability to supply food has increased as the proportion of spending on food is declining. As the fruits of industrialisation have been applied to the agricultural industry so a more scientific and technological agriculture has emerged. Yields have been increased substantially by the application of science. More and more food is available that no one in the developed world wants.

It is difficult to reconcile the embarrassing quantities of food in the developed world with the horrific starvation in many parts of the underdeveloped world. It is scarcely surprising that letters in the farming press should expose these thoughts to public scrutiny. Unfortunately, using food surpluses as aid in other than desperate circumstances may do more harm than good to the developing agricultural sector in less developed countries. Domestic agriculture in these countries may be undermined by food aid. Furthermore, the food aid that is being sent is, in many cases, rather expensive aid.

DEMAND FACTORS

Behind the income problems that have arisen, the inexorable operation of economic laws can be observed. Policy measures can have some effects on the way in which these laws operate but the agricultural industry has tended to forget about economic laws under the relative security of 40 years of government assistance. It has done so at its peril. For instance, over time, and for a variety of reasons, the demand for different types of food is likely to change. For many farmers these changes will be threats to their livelihood. To continue to grow products for which demand is declining will merely hasten the day when the support system must be overhauled radically.

When reasons for changes in demand are sought it is usual to start by exploring the effect of any price changes. There are major differences between one product and another. The consumer is far more responsive to a price change in beef than to one in eggs. Only when the producer is involved in direct marketing of unsupported products can the implications of these consumer responses be immediately felt. However, the same influences underpin changes in demand for all agricultural products and are especially relevant where prices are pushed artificially high by policy measures.

Secondly, the pattern of demand alters as a result of changes in the level of disposable income. As real incomes rise, so the demand for items like potatoes or cereal products for human consumption will tend to decline relative to the demand for more exotic foodstuffs. The implications of this are clear. For the producer of high-value exotic foodstuffs, the future is relatively healthy, whereas for the producer of low-value products it is more bleak so long as real incomes continue to rise. A second consideration when looking at the effects of income increases on food consumption must be that there is a demand for more prepared foods such as the mixed packs of washed and chopped vegetables found on the supermarket shelf. The benefits from adding value to vegetable products in this way are unlikely to be of much significance to the grower.

The changes in demand for one product can affect the demand for another product. The increased demand for polyunsaturated margarine has obviously had adverse effects on the demand for butter. This inter-relationship is likely to be a function of changes in taste as well as price changes, but whatever the cause it is very significant for the dairy industry.

Policies have been used to paper over the cracks and to generate additional supplies of high-cost products. The cushioning of the producer from the economic consequences of his actions cannot proceed forever. The dairy industry came to terms with these unpleasant facts of life in the turbulent years of the mid 1980s, as did the cereals farmer in the late 1980s. Coming to terms with changes in demand is a problem for the farmer as well as the agricultural policy maker. Changes in demand should guide changes at industry and farm levels. It is no longer reasonable to pretend to live in an unchanging world and then cry out for help as the waves of change beat against an ill-adapted industry.

SUPPLY FACTORS

It is the interaction of supply factors with demand factors that has made the problems of the agricultural industry so severe. At the time when demand levels for many products have stagnated or declined, supplies have increased dramatically. The ability of the industry to respond to the wartime and postwar call for increased production has been its downfall. This call was understandable in 1945 but rather less defensible in the 1970s as selective expansion of the industry was still encouraged in the face of European surplus. Milk was seen as a suitable candidate for expansion as the UK jockeyed for domestic advantage in the EC. Before long, doubts were being expressed about the efficiency of British agriculture in general[1] and dairying in particular.[2] Beneath the superficial appearances of technical dynamism and physical output increases there were signs that, in other ways, domestic agriculture was not in such a healthy state.

A major influence on the increases in supply has been the technological advances of the last 40 years. Animal and plant scientists in research institutes and universities have raised yields. Advisers have preached the gospel of increased production and the farming industry has responded. Average yields changed little between 1800 and 1940 but, after 1940, there were significant changes. The average yield of wheat more than doubled from 1940 to 1980 to almost 6 t/ha. Recent trials in research institutes have produced yields of 12 t/ha. Less dramatic increases affected other crops and livestock output. Living in the security of the postwar support system it would have been irrational for farmers not to have taken advantage of these advances. Prior to entry into the EC, levels of domestic self-sufficiency were increasing, but it was the entry into Europe that created a new perspective: self-sufficiency in the EC. The potential of the EC to over-supply itself with food has become a major embarrassment and a cost. Now that Europe is the frame of reference rather than the UK, these surpluses cannot be ignored.

The increases in supply have taken place in an economic environment that has not been wholly favourable to the farmer, who, like many others, has been the victim of a cost–price squeeze. With monotonous regularity the annual white paper on the industry has announced product prices rising more slowly than input prices. Technological improvement, increased output and increased efficiency of resource use have shielded all but the inefficient and the over-borrowed from the cost–price squeeze. However real the cost–price squeeze has been, it has been less brutal an influence on farm business adjustment than the hidden hand of the market would have been. The cost–price squeeze is likely to increase as the policies are adjusted to take account of growing surplus problems and increased costs of support.

Under the cover of 40 years of almost unparalleled agricultural prosperity, economic changes of great significance have occurred. These remained largely unnoticed by farmers even if they created nagging doubts for policy makers. Underlying economic laws were ignored and a policy of benevolent expediency followed. The industry responded by thinking more and more in terms of output and less and less in terms of economic efficiency. This production orientation has become more marked over time. Performance of enterprises is measured in physical, not financial, terms. The cereal farmer measures his returns in tonnes per hectare and strives for membership of the Ten Tonne Club. Such were the symptoms of an industry detached from economic reality and indicating that farming was, in Richard Body's phrase, 'in the clouds'.[3]

I do not wish to advocate the free market with quite the enthusiasm of Richard Body. The problem is not agricultural policy in itself but a particular agricultural policy which has created additional needs for adjustment in the industry. Free market medicine is a cheap opting-out rather than a real solution. It is possible to conceive of policies to soften the pain of economic adjustment, but to come to terms with why such policies have so far been so little implemented, the idiosyncrasies of the CAP need to be understood.

THE PROBLEMS GENERATED BY THE CAP

The CAP is both a reflection and a cause of the problems with which the agricultural industry finds itself involved. It is a reflection because it mirrors the problems of agriculture in the face of a declining demand for food products relative to other goods and services, coupled with the tendency of technology to push output up. It is a cause of contemporary agricultural problems in that the policy mechanisms have exacerbated difficulties instead of resolving them.

Under the deficiency payments system which was the main element of support prior to entry into the EC, there were certain advantages. The most important of these was the fact that the deficiency payment was a topping-up payment. Broadly speaking, prices were determined in the market place and government supplemented the market price with a deficiency payment. Prices to the consumer were thus significantly lower and, consequently, the amount consumed was greater. The consumer also had the benefit of relatively cheap food imports. A deficiency payments system is most rational where domestic production is a relatively small proportion of the total, for it remains relatively cheap to implement while allowing imports of low-cost food from world markets.

The relative openness of UK trading policy with regard to food was being eroded before entry into Europe. The policy of the EC dealt the final blow to the relatively free trade in food products. European farmers and politicians had favoured protectionism for agricultural production when the UK had largely ditched such policies in the nineteenth century. The consequence was the replacement of deficiency payments by a system of guaranteed prices and import levies. The precise policy varies from product to product but the general effect remains the same. The system creates an additional burden for the consumer as food prices are raised by the higher prices paid for farm products. Cheap imports are met with tariff barriers and the system passes the burden of support much more to the consumer. In contrast, the deficiency payments system was supported by indirect means. Income tax which could be described as progressive (those that can pay more do [or should] pay more) whereas a surcharge on food prices hits the poorer groups more and can thus be described as regressive.

The deficiency payments system was attuned to British conditions. The CAP never was. It was, and remains, a product of the political climate of Europe. The scope for change, even where there are glaring faults, is limited in the words of a senior administrator who has worked in the European bureaucracy: 'by the resistance of vested interests in conjunction with the egoism of politicians seeking re-election.'[4] It is not only the policy mechanisms that are at fault but also the decision-making processes that create the policies.

The result of the CAP has been an increase in output which has generated the surpluses in almost all major products. There is no rational mechanism to clear the glutted market. Prices remain high within the EC and surpluses are dumped on other countries at a cost to the Community. The cost of these surpluses is highest where the differences between world prices and EC prices are greatest. For this reason it is scarcely surprising that the first attempts at reform should have been directed against the dairy sector. The levels of surplus must be considered at an EC, and not a national level and, as Table 2.1 shows, there are few major products which are not in surplus, and many of the surpluses have increased rapidly in recent years.

The cost of the growing surpluses both to the consumer as tax-payer and to the government has been considerable. The gap between levels of direct support and incomes to farmers has narrowed and in one year, 1983/84, the direct support level exceeded the total farm income of UK farmers (Figure 2.1). In addition to the direct costs there are the hidden costs of higher food prices which are much more difficult to estimate, but which are likely to exceed the direct costs.

Table 2.1 Levels of self-sufficiency in agricultural products
in the then Community of Ten

	1973/74	1983/84	1985/86
		(% self-sufficiency)	
Total cereals (excluding rice)	89.9	105.4	113.7
Common wheat	105.7	117.3	120.8
White sugar	90.6	118.1	137.0
Beef and veal	99.9	112.0	105.6
Sheep- and goatmeat	66.6*	79.7	78.7
Pigmeat	102.0	102.0‡	102.0
Poultry	103.0	107.0	105.3‡
Skim milk powder	163.8*	101.4†	133.3†
Butter	110.4*	124.4†	132.2†

* EEC 9.
† Excluding imports under trade agreements.
‡ Estimate.

Source: Consumers and the CAP (1988)

The early attempts at CAP reform have been predictably unpredictable. The products receiving most attention have been milk products and cereals. The introduction of milk quotas in 1984 began to reduce the cost of the single most costly item in the EC budget. In 1986, quotas were tightened further. With hindsight it is possible to see three principal responses to quotas. The first, which was often an irrational panic, led people to quit dairying at a time when the salvage value of their stock and equipment was very low. The second, a more stoical response, was to tighten the belt and look for savings, particularly in feed costs. The third response was to diversify, and some of those who began to do so at that time have enthusiastically adopted and developed alternative enterprises.[5]

The cereals surpluses have been tackled differently, namely by the introduction of co-responsibility levies paid by producers when

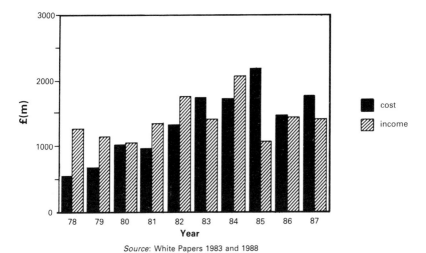

Source: White Papers 1983 and 1988

Figure 2.1 Incomes and direct support costs of UK farming 1978–88.

production targets are exceeded, and by the adoption of more stringent standards of intervention. A further measure designed to take land out of cereals is set-aside, which was introduced by the EC in 1988 and promoted enthusiastically by the UK. This scheme has important implications for the development of alternative enterprises and, like the US schemes, can reasonably be regarded as a form of diversification itself.

It remains unclear as to exactly how the specification of guaranteed quantities will interact with set-aside policies. If price reductions begin to exert any effect on production, then the marginal producers will drop out. There may be a case for social support to those marginal producers but the early signs of high levels of uptake in East Anglia do not suggest that problem farmers have been the beneficiaries.

The final element in the EC reforms, which embraces and extends similar principles to those evident in the US Conservation Reserve Program of 1985, is the establishment of Environmentally Sensitive Areas (ESAs). These offer farmers financial inducements to farm in an environmentally sensitive way. 'Environmentally sensitive' ways can include practices which retain archaeological remains, low-input farming, species-rich pastures or whatever is deemed appropriate in particular regions. The principle is not one of compensation but one of public payment for public good. There are ESAs from West Penwith in Cornwall to the Grampians and a wide range of sensitive and valued environments are included. The concept and its implementation are not without problems, but the principle of paying for environmental products represents an important development in policy and a type of alternative enterprise for farmers to consider in the designated areas.

Consumers have become more aware of agricultural policy since entry into Europe. The awareness has grown because the inefficiences and inconsistencies of the CAP have increasingly been exposed to public scrutiny. The public of the UK is largely a public of food consumers and not food producers. These criticisms have been aired thoroughly and critically appraised in a major report from the National Consumer Council.[6] Dumping of European surpluses, sometimes undercutting domestic production in the remoter parts of the world, and the ritualistic bickering at annual price determinations, have done little to enhance the public's conception of the CAP. It is seen as a cumbersome and ill-adapted set of policy instruments that are having not only a detrimental effect on its pocket but also on other matters that are dear to its heart – such as the state of the British countryside – and these wider issues are receiving increasing attention.

ENVIRONMENTAL PRESSURES

The rural environment provides the nation with more than food: it offers recreational opportunities, provides a backcloth of landscape to these recreation activities, contains wildlife and wildlife habitats which are of increasing interest to the wider population and, finally, provides other primary products such as water and timber. It has been argued with increasing legitimacy that many of these have been subjected to growing threats in recent years. The demands on land change over time and it is unfortunate that it has taken so long for the winds of change to be recognised. These demands for non-agricultural products have grown and cannot be dismissed lightly.

Access has long been an emotive issue amongst both farmers and recreationists. The divergence of opinion between recreational pressure groups who see access into the working countryside as a townsman's right, and the farmer who wants to protect his crops and farmland, remains considerable. Marion Shoard's book *The Theft of the Countryside*[7] created a great deal of debate which illustrated the strength of feeling on both sides. Furthermore, access remains an important issue on the Countryside Commission's agenda for action.

Plate 2.1   *Landscape in decline*
Stone walls and hedges have been removed; isolated trees linger
on as relics from a richer landscape. Not the 'prairie wastes' of
East Anglia but South Devon.

Much of the debate hinges around the nature and use of the
public footpath system. Many farmers regard the footpath system
as an anachronism. It is there because of pre-industrial needs when
most of the population moved on foot. Field enlargement, increasing amounts of arable land and threats to crops and stock have led
some farmers to divert, obstruct or damage the footpaths. The
instances of threatened footpaths are perhaps fewer than some
ramblers claim and rather more than the farming community will
admit to. To the ramblers, the footpath system is a licence for access
into otherwise inaccessible areas and footpath systems are jealously
protected. The refusal of the ramblers to agree sensible adjustments
to the footpath network has aroused the antipathy of farmers. Such
skirmishes between opposing camps are inevitable as demand grows
for different uses of the countryside. The debate about access must
be underpinned by the recognition that walking is, and is likely to
remain, the most popular active recreational pursuit outside the
home.
    The call for access is a call to get into a countryside that has

attractions. Whether it is the rambler walking in the landscape or the pleasure driver viewing it through his car windscreen, the scenery is undoubtedly an important contributory factor to enjoyment. Agricultural changes have wrought great changes on the landscape, particularly in the last 40 years in the predominantly arable areas of the country. The majority of the population feel that the removal of hedgerows, copses, ponds and hedgerow trees has resulted in a decline in landscape quality. What was condoned in war years as necessary to produce food is condemned in the 1980s as the surpluses build up.

One response of the farming community to criticisms of landscape change is to assert that landscape has always been dynamic and never static. This is perfectly true but irrelevant. If the population as a whole believes that the landscape changes that have occurred are unacceptable then farmers or agricultural policy makers (or both) must recognise that they are out of step and out of touch. Although landscape is normally only a by-product of other land-using activities it is an increasingly important by-product. The landscape effects of contemporary agricultural policy and farmers' responses are legitimate concerns that can no longer be dismissed. Landscape has become more important as the population has become more affluent and more mobile. As the means have been offered to the public to reach the landscape so it has been transformed before them, not everywhere to the same degree but sufficiently in some places to cause resentment and alarm.

As landscape diversity has declined, so ecological diversity has declined, too. In a decade which has witnessed such phenomenal growth in conservation organisations such as the Royal Society for the Protection of Birds, ecological value and conservation interests can no longer be treated lightly. The heated public debate which surrounded the passing of the Wildlife and Countryside Act in 1981 is evidence of the potency of conservation as an issue. Unquestionably modern agricultural methods have had adverse effects on wildlife. As in the case of landscape changes, so ecological changes have been concentrated in certain areas, but the changes have been sufficient to generate national concern.

There is more to the concern of many ecologists than mere species losses. The present practices of farming appear to be disrupting ecosystems at a deeper level. The increasing use of pesticides and herbicides and the substitution of inorganic nitrogen have created both income for farmers and problems for society as a whole. The lack of selectivity of many pesticides leads to the loss of beneficial insects such as pollinators and predators. The work being carried out by the Gamebirds Research Project has shown the degree of damage to wildlife in general and gamebirds in particular, caused by the insecticidal effects of certain fungicides.[8] The concentration of organo-chlorine chemicals at higher levels in the food chains of wild animals was well documented and led to a partial ban on their use. There remain major doubts about certain herbicides and pesticides, and the public has become increasingly critical of spray drift. The rise in nitrate levels in water supplies obviously is related to changing agricultural practices although the precise relationships remain imperfectly understood.

The nitrate levels in many East Anglian rivers now frequently exceed the recommended EC standards and the treatment costs will soon rise to millions of pounds. The price being paid for the environmental pollution generated by the agricultural industry is growing. It is not just the general public that is showing concern; so, too, are the water authorities and doubts have been expressed by the Royal Commission on Environmental Pollution in its recent reports. The privatisation of water authorities has given a great deal of publicity to the water industry and water pollution has been high on the agenda. Groundwater protection zones are planned where nitrogen applications will be limited or banned. Attempts to resolve the nitrate problem in water will be a major environmental issue in the 1990s. As contemporary farming practices throw up new problems for the chemical companies to solve, the ghost of good husbandry looks on and smiles ironically.

FOOD AND HEALTH

The relationship between food and health manifests itself in various ways: in malnutrition and increased disease susceptibility in the third world, and in the growth of diseases associated with overeating in the developed world. There is a growing public interest in these relationships and this interest is reflected in changing eating habits and attitudes. There are two major areas of concern. Firstly, the effects on health of certain widely consumed components of our national diet have been explored. Secondly, there is growing concern about additives, residues and contaminants in food, at least some of which may find their way into food products before they leave the farm gate. The furore over *Salmonella* in eggs, which led to much antagonism between MAFF and the Department of Health, is an example of this type of problem.

Over recent years the farming and wider press has been awash with articles concerning the relationship between animal fats and health. There is a strong body of opinion within the medical profession that links high levels of animal fat intake with increased risk of heart disease. Recommendations from major committees inquiring into the medical aspects of food policy to cut levels of animal fat intake cannot be ignored as an influence on eating habits and food policy. Meat and milk producers cannot pretend that changes have not taken place. Changes in milk pricing and greater penalties for fatty carcases are evidences of the growing concern.

The debate about diet and health is complex, but a consensus has emerged which indicates that a diet rich in saturated fats, sugar and salt is not nutrionally advisable. The critics of the food policy reformers have argued that it is the genetic differences in susceptibility to adverse health effects from diet that are the key. That there are differences is accepted by the reformers who still are prepared to argue the case for dietary reform. A national policy, it is argued, must set the guidelines for individual responses.[9]

The debate about fibre in the diet is of less significance to the farmer than the food processor. It indicates that the discussion about food and health goes beyond the single issue of animal fats. As tastes change in response to assertions or innuendos from the medical profession, the effects must be felt in the food production industry.

The wider public concern about food and health is reflected in the increasing consciousness of additives and residues in food. Public interest in additive-free food appears to be growing. Free-range eggs command a significant premium and a wide public following. The widespread use of hormone influencing growth promoters may bolster profits in the short term and yet raise doubts in the longer term. In the future, standards for use of agrochemicals, whether they are pesticides, herbicides or growth promoters, are likely to become more stringent. The notification of the banning of certain growth promoters in 1986 shows that consumers' attitudes can and do change agricultural practices. This has had knock-on effects and has been the immediate cause of acrimonious exchanges between the US and the EC (see Chapter 1). In this case, consumer doubts have overridden the absence of any clearcut scientific evidence to force the change. The consequence may be that organic or semiorganic premiums are eroded but those that have learned to live with self-imposed controls may be in a healthy position to develop their enterprises under the new regimes.

The concern with the quality of food products and the link with farming practices has been given a major public airing in the debate about *Salmonella* in eggs. The studiously qualified reassurances from the Minister of Agriculture have done little to quell public disquiet about the level of contamination. Furthermore, the public exposure of intensive farming practices has not enhanced the image of an already tarnished industry. Another type of contamination causing concern is the pollution of water by nitrates and other chemicals. The water issue is potentially of much greater importance than the eggs debate. Both are illustrations of the same problem: a failure of producers to see the consequences of their actions in the wider context. The general public is becoming increasingly critical of the tunnel vision of producers and government ministries that do not always appear to be operating in the interests of the public at large.

There are many reasons why the farming industry should be concerned about its prospects. The inevitable stresses on farm incomes in a developed economy are likely to be exacerbated by the illogicality of the CAP and compounded by an increasingly critical public that will no longer tolerate habitat destruction and landscape decay. It is not sufficient to attribute these changing attitudes to a hostile and intemperate press and it is clear that there is a growing estrangement between farmer and consumer. But there is a need to respond to these problems rather than to retreat behind stale and worn-out arguments. Part of this response should be an examination of the wider opportunities that farmland offers as a resource for producing more than surpluses of conventional foods. Policy changes will be needed if the farming industry is to adapt effectively to meet the needs of contemporary society, but the individual farmer's actions can lead the way. Alternative enterprises on farmland constitute one possible channel of escape.

New policy initiatives have appeared in the UK in the late 1980s. Some have emanated from Brussels such as ESAs and set-aside. But the response in the UK has been greater. The 1986 Agriculture Act broadened the scope of MAFF's work and the elements of the Alternative Land Use and the Rural Economy (ALURE) package provided some policy initiatives to fit into the broadened framework. Development on agricultural land was made potentially easier by the removal of MAFF's vetting powers on certain grades of land. Forestry and woodland were to be encouraged by specific farm-based schemes and diversification grants were to be paid both for assessing the feasibility of proposals and the development of enterprises.

The search for alternatives must go beyond the confines of conventional agricultural adjustment. Juggling between corn and horn will not suffice. Farmers can look to a host of possibilities from the set of resources that comprises the farm. Buildings or land which contribute little or nothing to agricultural production can possibly be developed for non-agricultural enterprises. There may also be a case for changes of use to alternative enterprises on what is regarded as farmland of reasonable quality.

The ideal alternative from the farmer's point of view should be a product or service that suffers from as few of the weaknesses of conventional agricultural products as possible. In summary the alternative should:

- Be capable of showing a return on capital that will match alternative uses of the same resources in the present and in the future.
- Be capable of experiencing a growth in demand (or likely future growth) at constant real prices.
- Be subject to natural restraints in supply expansion and satisfy only part of the national or EC demand for the product.
- Be independent of policy support mechanisms that might change to the disadvantage of the producer.
- Be environmentally acceptable by imposing no additional environmental costs.
- Be associated with good health and free from chemical contamination and any other health risks.
- Raise no consumer doubts about animal welfare.

In addition, from the national standpoint, any product or service which can contribute to export earnings is likely to be regarded favourably. In the future, employment creation could become a greater political issue and employment-generating products might receive more attention.

The search for the ideal product or service has begun. Sometimes it lacks direction, like the blind man's search for a black cat in a darkened room. Alternatively, the search may be planned, recognising a need to understand marketing principles, a need to review the options and a need to respond rationally to the challenge that is being posed.

REFERENCES

1. CENTRE FOR AGRICULTURAL STRATEGY (1980) *The Efficiency of British Agriculture*. Report No. 7. CAS.
2. AMIES, S. J. (1983) *A Comparison of European Dairying*. Report No. 37, Milk Marketing Board.
3. BODY, R. (1984) *Farming in the Clouds*. Temple Smith.

4. TRACEY, M. (1984) 'Issues of agricultural policy in a historical perspective', *Journal of Agricultural Economics*, **35**, 3, 307–18.
5. *THE INDEPENDENT* (1988) 5 December.
6. NATIONAL CONSUMER COUNCIL (1988) *Consumers and the CAP*. HMSO.
7. SHOARD M. (1980) *The Theft of the Countryside*. Temple Smith.
8. GAME CONSERVANCY (1985) *Third Newsletter of the Cereals and Gamebirds Research Project*. GC.
9. JAMES, P. (1988) 'Dietary reform: an individual or national response', *Journal of the Royal Society of Arts*, **136**, 5382, 373–82.

Chapter 3

HOW TO THINK ABOUT ALTERNATIVES

This chapter reviews alternative approaches to farm business management and advocates a much stronger infusion of the principles of marketing into agricultural businesses, especially into those operating alternative enterprises. The farming community is in danger of becoming ensnared by attitudes formed in different times, when food was short, policy support guaranteed and marketing an almost irrelevant activity that went on somewhere beyond the farm gate.

The history of alternative enterprises in British farming contain, at first sight, little to interest the advocate of marketing ideas or the progressive farmer. Those who developed alternative enterprises were seen as harmless eccentrics or failed farmers, neither of whom was held in especially high esteem by the farming community. It is easy to see how these impressions arose. Rarely could the uncertainty of the free market for unconventional alternative products be preferable to the certainty of price support offered for conventional products. No doubt amongst the ranks of those running alternative enterprises there were those who had failed at conventional farming and eccentrics whose ideas and hobbies had become a consuming passion. In addition, there was a third group, an object of sympathy rather than derision, consisting of bona fide farmers who had been forced to retire from more physically demanding tasks through ill health. A dismissive attitude to alternative enterprises and their operators lingers on in the farming community and clouds the recognition of a fourth and growing group amongst those who run alternative enterprises on farmland. This group is made up of neither failures, eccentrics nor invalids but those who have adopted what might be termed 'the marketing approach'. Its members have reviewed the markets for traditional agricultural products and perceived threats. They have reviewed the market for alternative products from farmland and seen signs of hope. Some have established profitable alternative enterprises. These people are the standard-bearers of a new approach which recognises consumer demands and responds accordingly. They are guided by the marketing approach which has been neglected for far too long by the farming community.

It is not just British farmers who are production oriented. The very nature of farming operations and the structure of the industry, with so many individual producers, makes a production orientation the norm. Wherever intervention measures exist to clear surpluses, the production orientation is likely to be magnified. Farmers in the US and New Zealand have been as likely to exhibit a production orientation as farmers in the UK. The painful transformation from production to marketing orientation evidenced in New Zealand is an indication of the importance of marketing as more free market policies are implemented. The farmers' markets of towns in the US, where farmers sell only their own products direct to the final consumer, are also a powerful reminder to the producer of the principle of consumer sovereignty.

A shift in the direction of a marketing approach by British farmers requires more than a small change in their attitudes to business management. It is needed by both farmers who are operating or contemplating alternative enterprises and those who operate conventional farm businesses. The business orientation of most farmers in the UK has long been towards production rather than marketing. This production orientation makes sense when support policies are geared to increasing food output and there are deficiency payments or guaranteed prices to pay the farmer for what the market does not want. However, in an era of surpluses of main food products, the production orientation is sustained illogically. The striving of individual farmers to produce more may yield them additional income in the short term but the aggregate supply merely adds to a saturated market. The consequence in a European economy where demand for food is almost static is to lower the market clearing price and raise the cost of supporting farm incomes. The emphasis of so many farmers on increased output rather than increased efficiency thus compounds the problems of surplus and rising support costs.

There are signs that recognition is growing amongst farmers of a need to get closer to consumers. This theme was much in evidence at the Oxford Farming Conference in 1989. A number of speakers felt that consumers had become increasingly alienated from farming practices with adverse effects on consumption. The most recent and potent example of this had been the headline-making debate about *Salmonella* in eggs late in 1988. One of the ways that bridges can be built is to bring the consumer to the farm for recreation or value added enterprises. Thus, the value of these enterprises may be greater than the financial rewards to the individual. Unfortunately, impoverished farmers – and there are many in the late 1980s – need more than a warm glow of satisfaction to keep them in business.

The production orientation not only manifests itself in farmers striving for wheat yields of 10 t/ha or for milk yields of 6000 litres/cow. Farmers must also beware of retaining farming systems simply through a combination of tradition, fatalism and habit. Some farmers retain unprofitable enterprises or reject the possibility of potentially profitable enterprises on entirely irrational grounds. The affection for black cattle in beef production or the refusal to countenance sheep on a farm are indications of a production orientation, but it is a production orientation rooted in tradition and prejudice, rather than one geared to producing more output.

It has been argued that the production orientation is only beneficial when there is a sellers' market and that 'this situation rarely arises'[1] in the farming industry. However, where agricultural product prices are sustained at levels far in excess of world prices for these products (whether or not there is a market for the products) it would not be unreasonable to describe the situation as a sellers' market. But this cannot continue indefinitely, and the inevitable changes in policy will require an orientation which replaces fatalism with foresight.

The marketing approach to agricultural production should not consist of last-ditch attempts by marketing 'experts' to shift food surpluses to unwilling consumers. It is not just the responsibility of institutions like the Milk Marketing Board or Food from Britain. It is an approach that can be contemplated and adopted by all farmers and it is especially important for farmers involved in alternative enterprises. As an approach, it is rooted in the recognition that the customers' wants and needs must be responded to. Production decisions should be based on this rather than on attachment to tradition.

There is nothing new or original about the underlying principles of marketing. They are essentially derived from economic theory where the consumer is seen as king and producers and distributors generate income and profit in their attempts to satisfy the demands of consumers. Although marketing principles have been in evidence for a long time, marketing techniques derive from disciplines such as psychology as well as from economics. Marketing principles provide the framework and marketing techniques provide a tool kit with which the marketing experts endeavour to keep the machine of the market economy functioning smoothly.

In spite of the apparently useful task identified as the realm of marketing experts, some would argue that there is also a less reputable side to marketing. The critics tend to focus on advertising and suggest that massive expenditure on advertising can hardly be in the best interests of the consumer. He, after all, must pay for the advertising as well as the product. The line between informative advertising and persuasive advertising, which some view as the boundary between respectability and disrepute, is blurred. Even informative advertising has been described as 'a campaign of exaggeration, half-truths, intended ambiguities, direct lies and general deception'.[2] Casual observation and the findings of appeals made to the Advertising Standards Authority remind us that these criticisms are not without some foundation. Neither is the deception wholly the domain of big business. The oranges and bananas sold from 'farm-fresh' farm gate sales and the 'close to beaches and the sea' of some rather inland farm holiday accommodation illustrate that the farming community is not entirely immune from criticism.

It would be an exaggeration to say that all advertising is evil and misleading and that all marketing is a conspiracy against the consumer. The study of marketing principles and the application of marketing techniques are fundamental considerations if the farming industry is to adapt and survive in the difficult years ahead. We must remain aware of both the potential and the problems of the marketing approach.

MARKETING PRINCIPLES AND AGRICULTURE

The study of marketing problems within the agricultural industry might initially appear to have little in common with the marketing approach being advocated. There has been a tendency in the past to focus on generalised issues rather than solving individual farmers' marketing problems but this is understandable in view of the structure of the industry. These applications of marketing principles to the study of the agricultural sector are by no means irrelevant. They provide the context in which the individual must work to resolve his marketing problems by the application of appropriate marketing techniques. The study of marketing at this generalised level has examined the way in which agricultural markets work and assessed the efficiency of the marketing process. There has also been much interest in the special institutions such as the marketing boards, their functions and the efficiency with which they perform these functions.

The classical applications of marketing principles to the farming industry have looked at market structures, conduct and performance. The fact that there are so many producers of a single, largely undifferentiated product such as barley or sheepmeat, means that the market place rather than the individual producer determines the product price. The buyers of the farm products may be rather fewer in number and the buyer may be in a position of much greater strength than the farmer in many transactions. The emergence of contract farming and the exacting standards demanded by the increasingly powerful supermarket chains are further evidence that market strength is not to be found in the farming community.

Much of the investigative work of market structures, conduct and performance is outside the remit of the study. However, whether the farmer is producing pigs or potatoes or a new food product like pleurotus (an edible fungus), he must take account of the market structures and his strength or weakness as an operator in these markets. The dramatic increases in the market strength of supermarket chains, particularly in the big three (Sainsbury, Tesco and Isosceles), create obstacles and opportunities which must be recognised. For the skilled farmer, the exacting requirements of the supermarket buyer afford an opportunity to forge closer links with the consumer. In the case of alternative products, which might be found on the delicatessen counters in almost all food supermarkets and hypermarkets, the small producer may be unable to provide the continuity of supply and the tight control that the supermarket demands.

The strength of the supermarket multiples and the weakness of the farmer may depress the agricultural community. However, there has been another market trend which has created new opportunities for many farmers. The rapid expansion of 'pick your own' (PYO) establishments and farm shops is an indication that there are at least some consumers who value the freshness of the product and the experience of picking it.[3] The proliferation of this type of market outlet indicates a ray of hope for the producer. Maybe the clean and efficient supermarket has divorced its food products from their origins in the countryside and left a market gap which the farmer can fill. In spite of all the quality control of the supermarkets, there are grounds for scepticism. Can a swede in Marks and Spencer's, wrapped in cling film, really be worth two-and-a-half times the price of the swede in the town's retail market place 200 yards away? Can those hard green tomatoes with the flavour of raw green potatoes really be class 1? The supermarket is not infallible. Its fallibility may be the farmer's opportunity to get closer to the consumer with the food he produces. It may be necessary to do more than just sell at the farm gate. The consumers' whims and desires must be catered for and, if they are, significant new opportunities may present themselves. The investigation of the structure and performance of markets forms a backcloth of information for those who have acquired a marketing orientation.

The concept of the market niche enables the confrontation between the producer retailer and the huge supermarket chain to be put into perspective. A niche is a part of the market that has been unexplored by larger firms which a smaller firm can satisfy and retain. A niche might exist because of a geographical locality, a particular product expertise or particular services offered to customers. The result is the same: that smaller enterprises can find a niche in the total market place. The danger, of course, is that if the niche expands, the larger firms may want to expand into it and threaten the smaller firm. The extent to which major food producers use countryside imagery to promote their products indicates the potency of the idea of small-scale traditional food production and shows how what is a deceit on the part of major food companies can steal back some of the niche market.

A second area of interest to students of agricultural marketing has been the evaluation of the need for, and performance of, marketing boards and other agencies influencing the marketing of agricultural products. These range from formal marketing boards with extremely wide-ranging functions and monopoly powers, to agencies like Food from Britain.

Institutions involved in agricultural marketing are frequently criticised by farmers and it is important that the performance of these institutions is subjected to rational public scrutiny. It is debatable whether these institutions always benefit the producers but pertinent to note that where there are no official organisations, there are frequently demands for their creation. In the alternative enterprises area, three examples illustrate this point. Firstly, the Farm Holidays Bureau at Stoneleigh has been established to coordinate the marketing of farm holidays and to promote local marketing groups. Secondly, British Country Foods was formed to produce a directory of country foods and an association of producers. Its prospectus was prepared by the English Tourist Board, *Farmers Weekly*, the Institute of Grocery Distribution, National Farmers' Union Marketing and the Royal Agricultural Society of

England. A notable but characteristic omission from this list is the Ministry of Agriculture, Fisheries and Food (MAFF) and the Agricultural Development and Advisory Service (ADAS). Thirdly, there have been moves to establish an improved marketing system for organic foods, and to make organic standards clearer to the consumer. The presence of four organisations – the Soil Association, Organic Growers' Association, British Organic Farmers and Organic Farmers and Growers – has produced at least two sets of standards, but in 1985 three groups began to operate from under one roof in Bristol and this will hopefully ease the problems of standards and enable a more co-ordinated approach to marketing. National standards have been agreed in the UK in 1989, and these are likely to be followed by European standards. This should minimise the risk of rogue marketing.

Some of the institutions involved in marketing are purely marketing organisations. Others function in a wider sphere offering advice, stimulating production-related research or acting as producer pressure groups. They exist throughout the Western world. There is an equivalent organic organisational grouping in New Zealand to that in the UK. In the US, the existence of the Minnesota Wild Rice Council has already been noted. More recently, a North American Bramble Growers' Association has been founded. Farmers' organisations, including co-operatives, can benefit the membership as a whole by fulfilling a vital role in marketing and may also be able to contribute to the development of alternatives in additional ways.

It is undoubtedly difficult for the small man to market his product effectively. Agencies, whether public or private, have an important role to play. In the case of alternative enterprises in the UK, some of these agencies obtain the official blessing of MAFF and some do not. The value of this benediction is uncertain. What is more important is that the organisations that develop are not bureaucratically top heavy, that they are sensitive to consumer needs and that they provide producers with effective and accurate marketing information.

The third main area of study of marketing principles lies in the analysis of marketing aspects of the interaction of supply and demand. Thus, the changing demand for fats might be explored and the implications on UK milk producers considered in the light of the future demand-and-supply situation. These issues are unlikely to appear immediately accessible to the individual farmer and studies of this type will rarely be commissioned by individuals. The findings of such investigations are, though, vital ingredients for successful decision making. We live with the problems of over-production by farmers operating in a business sector that has become detached from economic reality. If the marketing approach is to be adopted widely in the farming community, the false world of intervention prices must be shunned in favour of the reality of the market place. Individual business decisions should be made in the light of analysis of future demand–supply interactions. Incomes can be enhanced greatly if products can be identified which have a growing market or if marketing efficiency can be improved.

The need to explore demand–supply interactions is even more important where the farmer is producing goods or services which are unsupported by CAP price arrangements. The wide range of alternatives and the lack of any significant institutional support for demand–supply studies of alternative products from farmland leads to decisions being made in ignorance and uncertainty. Reference can be made to some studies which may be relevant. The analysis within a consultant's report to the South West Economic Planning Council on Tourism in the West Country may yield estimates of future demand for tourism.[4] Individual operators might respond by expanding enterprises for which demand is likely to grow and alter their promotional strategy accordingly.

To summarise, the principles of marketing have been explored in a number of ways. At times, these investigations might appear rather remote from the individual producer. However, within these studies, certain principles of marketing are in evidence. The recognition of these principles is vital to the future welfare of the agricultural industry and even more critical to the welfare of that part of the agricultural industry which is unsupported by the CAP. Unless certain questions are asked and answered, the production of alternative goods or services will remain a lottery, with many losers and a few winners. These questions can be summarised as follows.

- What is the structure of firms in an industry or sector of an industry? How is the industry (sector) evolving and what are the implications for the marketing of its products?
- What institutions exist to influence the marketing of the products of the industry (sector)? Who are the gainers and losers from the activities of institutions? Are the institutions' activities benefiting the industry (sector) and could the benefits to consumer and producer be enhanced by additional or different institutional activity?
- What is the demand for a given product? How is it changing and how is it likely to change in the future? What are the implications of these demand changes on individual producers'

profitability? What strategies can be adopted to enhance producers' profits? What factors influence the supply of a product? Are there barriers to entry or can anyone easily begin to supply the product?

The study of these questions requires the application of the principles of marketing. They are ignored at the producer's peril.

MARKETING TECHNIQUES

The term 'marketing techniques' is normally applied to the range of specific techniques that an individual producer (or someone acting on his behalf) might use to improve the marketing of his products. These techniques are built on the foundation stones of marketing principles but focus on the factors within the marketing process that the producer has at least some control over. The concept of the *marketing mix* is frequently applied to the factors wholly, or partly, within the producers' control that can influence buyers' responses. Selecting the appropriate marketing mix is the task of the *marketing plan* which should also contain information on implementation and monitoring of the marketing process. The whole process is frequently termed 'market research', although on occasions the term is applied to the rather narrower range of marketing techniques associated with exploring the market for a new product. An integral part of the marketing process is an analysis of the market (Figure 3.1).

IDENTIFYING THE CUSTOMER

Fundamental to any marketing plan is the identification of the market, or markets, for the product. Marketing experts have given considerable attention to the range of social, psychological and economic factors that influence buyer behaviour.

At a general level, culture exerts a powerful influence on the range of goods and services that a society makes use of. The term 'culture' in this context means the set of beliefs, values, attitudes and habits that groups of people share. Cultural influences create collective preferences for particular types of food and particular types of leisure activity and landscape. In spite of some evidence of an emerging global culture, a visit to the Far East or continental Europe will rapidly remind one of the major cultural differences that remain.

The different products demanded by different cultures may

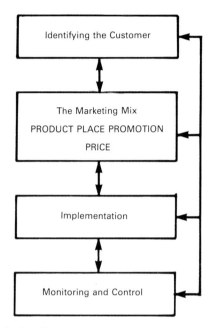

Figure 3.1 The marketing planning process

create opportunities for British producers, just as the UK has provided a significant market for French, Dutch and German cheese producers. Feta, a type of cheese widely consumed in the Middle East, has been produced in Scotland and exported to Iran and Turkey. Another Scottish attempt to identify the demands of overseas customers can be seen in the groundwork conducted in the Scottish Tourism and Recreation Studies of the 1970s. The demands of different cultures could be carefully related to the particular resources. Marketing and promotion abroad could focus on the qualities that different groups sought. Those groups with preferences for artistic resources might be directed towards Edinburgh and those with alcoholic preferences towards Speyside and its distilleries. A recognition of the forces that shape consumer demand can benefit the producer in many ways, from consumer goodwill to the realisation of new opportunities from domestic or overseas customers.

Within any country, there is likely to be considerable cultural variability. The term 'subculture' is applied to identifiable groups with distinctive differences of culture. These differences may be ethnic in origin or can be determined by social class. Thus, we can identify Asian or Scottish subcultures or working-class subcultures. The significance of subculture lies in the different demands for goods and services that different subcultures generate. The desire of ethnic minorities for goatmeat, for special slaughtering methods or for particular types of vegetables that can be readily produced under polythene in the UK offer opportunities to domestic producers and processors. Likewise, the different classes might be seeking different styles of holiday with different facilities and recreational opportunities.

In attempting to understand consumer behaviour at an individual level, marketing analysts have employed psychological concepts. What the consumer perceives may differ from what the seller intends. Those who are responsible for marketing must take account of how people's minds work and the nature of their perceptions. How people learn has been the subject of much debate in psychological circles. Whilst some credit Man with the learning behaviour of Pavlov's dogs, others suggest a more complex learning process based on the acquisition of more abstract mental guidelines. Both theories merit the marketing analyst's attention, for if food buying behaviour is routinised in terms of where and what people buy, the scope for new products in new outlets may be limited. However, if a more complex process of nutritional and culinary consciousness is developing, and this can be associated with the provision of new outlets like PYO or farm shops, or new exotic and health-inducing foods, a rather more optimistic future might be predicted for alternatives.

The consumer must not only be aware of a good or service. He or she must also be motivated enough to acquire it. He or she must *need* something. Maslow has identified a hierarchy of needs which begin with the physiological need for food and shelter, move through the need to belong and to be respected, to a final highest need for self-fulfilment. Much advertising directs itself to cater for certain needs for health and security whilst the dinner party camaraderie of meat advertisements in women's magazines play on the needs of consumers to belong to a genteel and respectable social milieu.

The extent to which psychological concepts can assist marketing studies has been debated. The relationship between personality and consumer behaviour is conflicting.[5] Some argue that personality is an important influence on buying behaviour whilst others have suggested that there is no clear relationship. Personality is an important ingredient in the construction of *psychographic profiles* which attempt to combine psychological, social and economic variables in an analysis of *life style* and allow the categorisation of consumers into different groups.

The objective of the detailed examination of consumers is to understand buying behaviour and to subdivide consumers into meaningful and relevant groups in relation to a product or service. This division of consumers into groups will include economic and social variables and is likely to include demographic factors such as age and family status. This process is termed 'market segmentation' and should be the culmination of the process of identifying the consumer.

What people comprise the important market segments for a farm museum in Somerset? Visitors may come by car or in organised groups on coaches; they may be very young (pre-school playgroup outings) or very old (old-age pensioners' outings); they may be farmers or rural people wanting to wallow in nostalgia or townees seeking a fun day out; they may be locals or tourists from other parts of the UK. The process of market segmentation should greatly aid the selection of the marketing mix by gearing the package on offer at the farm museum to the most important market segments. The segmentation process may well reveal three main segments, which can then, if necessary, be subdivided.

- Family groups (car owners).
- Organised parties (excluding educational).
- Educational visits.

The market segments for a PYO farm, for self-catering accommodation, or for firewood from farm woodlands are likely to be entirely different. The contrast in the market segments for firewood consumers illustrates the difference (see Figure 3.2).

The process of market segmentation is fundamental to effective marketing. The producer (or seller) must consider who his likely consumers are, where they are, how they will find out about the product and where they will acquire it. There is little point in trying to establish a farm shop to sell gourmet farm-produced cheese in mid Staffordshire when Harrods would be a far more appropriate sales outlet. The *Sun* may be an inappropriate newspaper in which to place advertisements for self-catering farm-based accommodation appealing to the discerning wilderness seeker.

Market segmentation is a necessary first step towards targeting, which is the strategy used to reach the chosen consumers.[6] An organisation may opt for the largest single market and pursue an *undifferentiated marketing* strategy to all potential consumers.

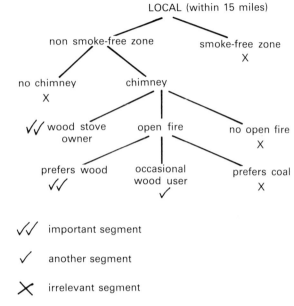

Figure 3.2 Identification of market segments for firewood

Alternatively, the organisation may pursue a single narrow segment with a *concentrated marketing* strategy. Finally, the organisation may pursue several different segments with different strategies and marketing mixes, which is termed *differentiated marketing*. When the consumers have been identified clearly, the marketing mix can be adapted to their needs.

The isolation of market segments and the development of an appropriate marketing strategy are activities which the producer of alternative products can and should indulge in. It is not necessary to be a multinational corporation to think in this way. The pony trekking establishment may cater for two different groups – local residents and tourists – and differentiate its marketing strategy accordingly.

THE MARKETING MIX

The marketing mix is normally considered by examining the four Ps that make it up: Product, Place, Promotion and Price. The selection of the appropriate marketing mix is embodied in the marketing plan. Each of the contributory factors must be examined and examples will be used from different types of alternative enterprise to illustrate specific points.

Product

Marketing textbooks stress the highly competitive nature of economic activity and the need for businesses constantly to generate new products within a particular field in order to survive. The idea of a product life is used to describe the infancy of a new product, the growth in sales that leads to maturity and the decline in sales associated with senility. The cereals farmer or the beef producer are operating in rather different worlds from the microcomputer manufacturer whose product may be outdated tomorrow by the next innovation. The difference is most extreme when a staple food is contrasted with a modern piece of technological gadgetry, but in the case of alternative enterprises, the concept of a product life may have more relevance. For example, the mid 1980s consensus was that good-quality self-catering accommodation on farms constituted a potential growth area. Yet was there certainty that the demand for such accommodation would last? The evidence of the late 1980s is more ambiguous. Will new styles of holidaymaking, perhaps in other parts of the world, offer alternative and more attractive ways to 'get away from it all'? Another example that might be considered is the case of farmhouse production of dairy products. As farmers ponder over new ways to shift milk surpluses, there is likely to be a demand for additional Milk Marketing Board licences to produce farmhouse cheese and an expansion of output of whole milk yogurt and ice-cream which as of 1989 have avoided the quota regulations. Those farmers already producing dairy products on the farm are likely to face intensifying competition which may threaten their product. Therefore, product lives for alternative farm products may be relatively short and the factors influencing the length of the product's life must constantly be borne in mind by the producer.

The conventional farm produces a relatively small range of products rarely sold to the consumer without further processing. In marketing jargon, the product mix is relatively narrow. There has been a tendency towards increased specialisation in the agricultural sector which has led to a decline in mixed farming with its wider product mix. Furthermore, the particular patterns of price support of the last 40 years may also have contributed to a narrowing of the product mix. Whilst economies of size may dictate specialisation into one, or at most a few enterprises, the perils of operating with a narrow product mix were forcefully brought home to the specialist intensive dairy farm with the introduction of quotas.

The range of products on a farm with alternative enterprises may be greater but in this case, there is a need for a degree of consistency in the products offered. Jumping into a wide range of unrelated alternative enterprises is fraught with difficulty. Some alternative enterprises may produce final products for sale at the farm gate, others may produce specialist foods for an international market, others may appeal to a specific clientele (such as riding establishments). It is imperative that the range of goods and services offered is complementary and that the product mix contains an underlying rationale.

Once the range of products has been identified, it is important to think about the characteristics of the product that is to be offered for sale. Thus, the *core product* might be farm tourism. The particular farmer (or farmer's wife) must decide upon the type of farm tourism and the quality and nature of the farm holiday experience to be offered. This is termed the 'formal product'. Finally, there may be certain ancillary services provided. The sum of any ancillary services such as self-guided trails around the farm, leaflets, etc. constitute the *augmented product*. The provider of the alternative enterprise must look carefully at the products that could be offered. The products must constitute a suitable product mix from the consumer's point of view, not a random assemblage of alternative products that the farm might just be able to put together. Over time, the product mix may need to be varied either to entice new customers or to adapt to changes in consumer demand.

The identification of the appropriate product mix and the answer to the question: What business am I really in?, is by no means a simple task. The rare-breeds farmer may believe that his task is to maintain rare breeds. Or is his task to display rare breeds to the public in such a way as to provide an enjoyable visitor experience? This can affect dramatically the way the business is run in that if the rare breed comes first, the rarity will receive great attention, whilst the areas like the pets' corner where children have free access to common pets and farm animals will be an unnecessary nuisance. If he is aiming to create a visitor-oriented business, the wants and needs of the visitor should not be neglected.

Place
The location of a farm-based alternative enterprise is constrained by the location of the farm, which historically has rarely been purchased with a view to its potential for non-farming enterprises.

However, the place of production is not necessarily the place where the product is bought by the final consumer. Where the product is farm tourism or PYO, the place of production and sale are identical, but with farm-produced foods like pâtés or cheeses there are many potential outlets, amongst which the producer should aim to select the most appropriate.

The constraints imposed by location are a fundamental and vital consideration in alternative enterprise production, particularly in those enterprises where the points of production and sale to the consumer are identical. It is not just a question of living in an area with tourist potential or with a potential for firewood sales or PYO. A new bypass can remove the passing trade from one location and transfer it to another site. The narrow lanes of Devon constrain the ability of touring caravanners to reach certain locations. Each location is a unique combination of features combining accessibility, landscape and other features. There will be vast differences in the potential of farms to offer alternative products for sale at the farm. To establish an enterprise in a sub-optimal location creates a range of problems, including low visitor numbers, high advertising costs, risk of competition from better located farms and a high risk of enterprise or business failure. In addition to the problems associated with a poor location, there is also the inevitable challenge of intense competition in highly favoured areas for such enterprises as PYO or tourism.

One very apt illustration of the importance of place is provided by barn conversions. Different areas have different planning constraints and differences in demand for residential conversions. A barn in a high-amenity area with high residential demand can be worth in excess of £100,000; a courtyard complex well over £500,000. In other areas the value of the barns is likely to be nothing more than their agricultural value.

In the case of goods and services produced but not sold on the farm, the decisions must be made about where to sell in terms of areas and outlets. The producer of goats' milk has to choose among farm gate sales, local health food shops, supermarket outlets, hospitals with allergy sufferers and ethnic minority markets in London and other large cities. Different outlets demand different quantities and different stocks. The organisation of the transportation of the product may pose difficulties where the outlets are scattered and the level of demand is low. The shelf life of products is an important consideration. One has only to look at the 'imported' vegetables on some of the outlying islands of Scotland to see the relationship between consumer appeal and product quality. The recent developments in polythene tunnel production in these areas illustrates what can be done to fill a market gap.

The choice of sales outlets must be based upon rational assessment and not the destination of a pin on to a list of possibilities. The wrong place of sale will lead to the product not being sold even though there may be a demand for it, which could be satisfied if the product were sold elsewhere. This is illustrated by the early history of Japanese wood mushroom production in Virginia in the US. When growers expanded production they found that saturation of local markets was causing price cutting. With adequate market intelligence they would have been able to shift large volumes of the product in Atlanta wholesale markets at very good prices. If selling points cannot be linked to the production place at a reasonable cost, there is no point in establishing alternative enterprises. Where the product is a new food there are existing organisations that might assist. These range from co-operatives, to private marketing agencies, to public sector developments like the recently launched Devon Fare, initially funded by Devon County Council, or the national Food from Britain.

Thus, whilst the place of sale of conventional agricultural products is often fixed, as in the case of milk sales to the Milk Marketing Boards, or chosen from a small range as in the case of livestock markets, the determination of the places of sale constitutes a major challenge to the person establishing and running an alternative enterprise. The skill is in understanding potential outlets, complemented by an understanding of the constraints and opportunities created by the particular location of the farm. Where the product is sold at a place away from the point of production, it will frequently be necessary to seek outside assistance from a public or private sector agency. The producer should not absolve himself from concern about the choice of retail outlets, for this can be a crucial determinant of the success of the product.

Promotion

Products require promotion in order to create consumer awareness and interest. It does not matter whether the product is a new food, farm tourism or a farm recreation enterprise. The potential customer must be made aware of the product. The promotional strategy is the means by which customer awareness is created. The awareness may then develop into interest in and purchase of the product. Promotional strategies will need to vary greatly from one product to another but an examination of examples reveals common strands which may be borne in mind.

The promotional strategy of a farm museum is likely to have a number of elements. Any manager of an alternative enterprise should endeavour to maximise the free publicity that comes from articles in the national or local press. New developments or interesting acquisitions should be reported and invitations offered to journalists and photographers to visit and see the site in a fully operational state. If the museum is located in a tourist area, the distribution of advertising material should be more than the dumping of bundles of leaflets at information caravans, hotels and guest houses. Complimentary tickets to proprietors and information officers may lead them to make a personal recommendation which is perhaps the best of all forms of promotion. The manager should identify the segments of his market and the promotional strategy must take the various segments into account. There may be a significant proportion of educational visits and, for these groups, the educational significance of the museum should be emphasised. Help to teachers in the form of teacher packs, or teacher evenings or weekends may generate increased use and more effective understanding of the museum. Effective promotion requires a careful consideration of the best means of reaching the potential visitors to a site.

Advertising is clearly an important part of the promotional strategy. In addition to the efforts that go into displays at shows or personal promotion and selling, the consumer must be reached by use of time or space paid for in the media. There are many ways of advertising and a wide range of potential media in which to advertise. The farm museum is likely to use posters and leaflets in holiday accommodation and information centres, advertisements in local, and perhaps regional, newspapers, and may also use local radio and regional commercial television. Roadside advertising may be desirable if it can be located suitably and planning permission obtained. There may be a case for membership of promotional groups which market a variety of attractions in a common brochure.

Some farmers are loath to spend money on advertising and frequently regard money paid to advertising agencies as a waste of scarce funds. However, if an apparently expensive advertisement generates £1 more profit than its total cost, it can be justified. It is vitally important that the advertising media are carefully chosen and that the response is monitored carefully. Agencies may offer much-needed assistance to the farmer in advertising his product. Amateur advertisements are frequently a liability but the employment of professional agencies is no guarantee of success.

The correct choice of advertising media and an appropriate advertising budget must be, to a certain extent, a matter of trial and error. The prospects for success can be enhanced by rational thought and the answering of some basic questions including the following.

- Who are the customers?
- Where do they live?
- What newspapers/magazines/trade journals do they read?
- What is their use of commercial radio and television?
- What is the cost of advertising in the different media?
- What was the advertising medium that stimulated the customer's response?

If these questions can be answered and the various market segments identified it should be possible to devise a promotional strategy that enhances the prospects of establishing or sustaining an alternative enterprise.

Price
The price that is set for an alternative product is unlikely to be determined by the auctioneer's hammer. Competition amongst producers of similar or identical products, especially in a slack market, will lead to price cutting as the laws of supply and demand work to establish a market clearing price. Conversely, the producer of a unique product experiencing a growth in demand may be in a strong position to charge higher prices.

Where a price must be set for a new product or an established product in a new location, the choice of an appropriate price poses more difficulties. A high price generates consumer resistance. A low price sows seeds of doubt about product quality. Guesswork does not always produce the right results. The case of Curworthy Cheese, the product of a *Farmers Weekly* farm, illustrates the point. A frank article in 1983[7] looked back on the mistakes and acknowledged that the quality of the product did not justify the price tag that had initially been put on it. In addition to a number of other adjustments, the price of the product was reduced by 30 per cent in an effort to boost sales and sustain this new enterprise.

Different types of alternative enterprise create differing degrees of difficulty in setting an appropriate price. In the case of farm tourism, it is relatively easy to set a price comparable to competitors. A new entrant might well set a price slightly below the average rate to stimulate additional custom, whereas an established enterprise might be able to charge more for the known quality and service that attracts return visitors. With day visitor attractions, it is worth considering the price of competing attractions even if they have nothing to do with farming. Within a complex product mix such as a major farm visitor centre, there may be a juggling of prices of the various components of that mix. There is unlikely to be a similar profit margin on all items within the mix and the margin should be determined in response to both product charac- teristics (e.g. short shelf life) and consumer attitudes. If the visiting public feels cheated and 'conned', the benefits of return visits and personal recommendations are lost. Generally, the price that the consumer initially sees should be the price that he pays for the product. Additional charges for particular parts of the product mix (e.g. rides on a farm trailer or baby-sitting in a farm bed-and- breakfast establishment) may be resented by the customer.

Pricing of alternative products is made more difficult by seasonal- ity: PYO products, farm tourism, visitor attractions or firewood are all products which have a marked seasonality of demand. Conse- quently, it may be desirable to adjust prices to ensure greater off- season demand. Where there are similar products on offer in the immediate vicinity, this may pose few problems but in other instances it may be necessary to resort to trial and error to deter- mine appropriate prices at different seasons.

Market window analysis is a technique used to explore production possibilities of horticultural alternatives.[8] A market window is an estimated time when the wholesale price of a commodity exceeds the variable and fixed costs[9] associated with its production, packag- ing, marketing and delivery to a specified market. The essential information needs for market window analysis are production cost data, information on all marketing costs, information on competitors' production and marketing costs and information on price variations in wholesale markets.

The method is only suitable where there are seasonal price vari- ations normally caused by different times at which products are ready for harvesting. It would be highly appropriate to such estab- lished crops as new potatoes and would be a useful analytical tool with, say, strawberry production outside the peak season, or with alternative crops like artichokes or horticultural products under polythene in the Northern or Western Isles of Scotland. It has been used to analyse the prospects for cantaloupe and broccoli production as alternative enterprises in Virginia, in the US.

The technique of market window analysis is just one illustration of the need to analyse the production and the marketing situations affecting the individual producer in the context of the market as a whole.

As well as taking account of seasonal changes in demand in setting prices there is a strong case for contemplating alternative prices for different market segments. This is most obviously the case where day visitor attractions charge different prices to educational parties. Equally, it may be beneficial to offer preferential rates for tourist accommodation to parties, those who book in advance or old age pensioners, especially when they can be accommodated at offpeak times. This principle of different prices to different customers may also be applied to alternative foodstuffs with top-quality supplies of eccentric foodstuffs commanding premium prices from London delicatessens, and low-grade products being sold more cheaply at the farm gate.

The Marketing Plan

Design
The idea of a marketing plan is to draw together the essential considerations of marketing into a planning document which will be used to guide an organisation's marketing activities, rather than gather dust in the office. In practice, marketing plans vary from organisation to organisation. Sometimes they embrace the whole business plan of a market oriented business; at other times they focus on a more narrowly defined set of marketing issues. In the early life of an enterprise it is all too easy to ignore the marketing issues and concentrate on making the production system work. Whilst this response is understandable, it is also dangerous.

There are a number of stages which can be considered important in any marketing plan. They can be listed in the form of the following questions.

- What is the product and what are the normal expectations of production, sales, etc?
- What are the strengths, weaknesses, opportunities and threats of the enterprise from a marketing perspective?
- What are the business/enterprise goals and objectives? Do they need modifying in the light of the preceding questions?
- What options, strategies and choices are available with regard to marketing? Which of these is most appropriate?
- What is the cost of the proposed action?
- How can the operation of the plan be monitored and corrective action be taken where necessary?

Analysis

It is not sufficient to consider the four Ps (Product, Place, Promotion and Price) in isolation. The system is greater than the sum of its parts and the elements that comprise the system must be drawn together in the marketing plan.

Analysis depends upon adequate data which are often lacking with alternative enterprises (Chapter 5). It is important to make best use of available data. These data should relate to the internal operations of the enterprise or firm and the external relations of the product and its market.

Implementation

The plan will detail the pathway to production and marketing. The two elements of production and marketing must be phased correctly. All too often the first visitors arrive when the paint is scarcely dry and occasionally when it has yet to be applied. Recently constructed self-catering chalets may be set in a sea of mud that resembles a building site. The launch of a new food product must not precede the elimination of teething problems associated with its production. The maintenance of quality of certain dairy products, especially cheeses, may create considerable difficulties at the product development stage but these must be resolved prior to the product's full public launch.

The marketing plan should not be an inflexible document to be adhered to rigidly. It should be sufficiently flexible to identify appropriate responses to foreseeable circumstances such as fuel price changes or poor weather. Furthermore, if unforeseen and unforeseeable events demand changes in the way the plan is implemented, then deviations from the plan should be accommodated and the plan revised. For example, a marketing plan for a PYO farm might contain provisions for the development of snack bars, picnic areas or farm trails at specific times. Local competition may necessitate an adjustment in the product mix to differentiate the farm from its competitors to retain the advantageous position of the initial producer.

The marketing process hinges around the recognition of consumer wants and needs and the pursuit of a series of rational steps to meet these needs and at the same time to create profits for the producer. The marketing plan provides the details of procedures to be followed to produce and market the product. If it cannot be implemented or is unable to offer the producer the required guidelines, the marketing process has failed.

Monitoring and Control

In a survey of over fifty operators of farm holiday accommodation in Devon in 1983, only a handful of operators made any attempt to assess objectively the success or failure of different advertising media used.[10] If this is indicative of a wider neglect of monitoring the effectiveness of the marketing strategy where farmers have developed alternative enterprises, it is a cause for serious concern.

There are many ways in which the producer can monitor his enterprise performance. Different market segments can be given market research questionnaires to fill in to find out their likes and dislikes. Visitor books or suggestion boxes may provide clues to likes and dislikes. It is not sufficient to complacently assume that consumer satisfaction can be subjectively assessed. Robert Burns' plea:

O wad some Pow'r the giftie gie us
To see oursels as others see us!

is vitally important in enterprises where farmer and visiting public meet face to face. The enterprises are catering for the needs of 'others' and how they see 'ourselves' must be considered by enterprise managers.

Farmers' inexperience of advertising may be a cause of their failure to monitor advertising effectiveness. Where different journals or newspapers are being used, it is possible to alter the initials of the farmer to indicate the medium being used. Alternatively, some other coding should indicate where the advertisement is placed. Where telephone bookings are taken, or when visitors are informally questioned, it should be relatively easy to ascertain where they found out about the product on offer.

Monitoring and control of the marketing process should be an integral part of the application of marketing techniques to any alternative enterprise on farmland. The feedback that comes from the monitoring allows any necessary adjustments to be made. In the absence of any monitoring, the causes of any weaknesses cannot be pinpointed. Remedial action in the absence of adequate feedback can only be based on hunches which are occasionally right and often wrong. Attention to monitoring the marketing process and making any necessary adjustments to the marketing plan constitute the final link in the marketing process.

THE NEW BUSINESS ENVIRONMENT IN THE COUNTRYSIDE

In Chapter 2, the contemporary problems facing the agricultural industry are reviewed and the economic, political and social forces creating changes in rural policy are examined. This chapter introduces the marketing approach to farm business activity. The adoption of a marketing approach should begin with a review of the markets for farm products.

It has been argued that 'declining industries have invariably failed to define their business in sufficiently broad terms'.[11] Few people doubt that the relative importance of the agricultural industry will decline in a developed economy. Forty years of attempts to revitalise the farming industry have led to a situation today where farmers are more indebted and more pessimistic than at any time since the Second World War. Although the future may not be as bleak as the most pessimistic forecasters suggest, a substantial restructuring of the farming industry seems inevitable. Part of this should include a redefinition of the field of business activity of farmers.

A primary reason for the adjustment of the farming industry has been its alienation from the market place. As consumers' demands have changed, so agricultural policy has stood still. Policy instruments alone cannot be seen as the scapegoat because the farming industry has long prided itself in its links with policy making. Farmers and their representative organisations have been an influence on policy and not just unwilling beneficiaries. No longer can farmers expect to be shielded from market forces by the policy mechanisms. In the words of John Taylor:

> Farmers must look beyond the farm gate and react to consumer demand. Consumers are the ones who keep farmers in business and will be the ones who ultimately decide how the countryside will look.[12]

The narrow definition of agricultural activity is no longer tenable in the eyes of the consumer. Food in abundance is being seen as only one element in a package of products that the countryside can offer. The pursuit of food in abundance has reduced the quantity and the quality of the other products that are being sought by consumers.

The new prospects can only be scanned if the blinkered vision is replaced by a broader view. The changes in the attitudes of the consumers must be explored and the alternatives should be appraised in the light of such considerations. The remainder of this chapter identifies the key areas of change and leads into a detailed appraisal of the main groups of alternatives.

Increased affluence has given the population of the UK the opportunity to disperse itself from the industrial centres for residential and recreational purposes. The evidence of preferences for living in more rural communities is seen in the movement of population into the countryside in every postwar census. In the early postwar years, the more remote rural communities were losing population but many of them have become gainers of population as the flight to the shires has continued in the 1970s and 1980s. The bulk of the population of 'rural' Britain no longer relies on the primary sector (farming, forestry, fishing and mining) for its welfare. Instead, it consists of a population of urbanites who prefer to live in the countryside rather than the towns and cities.[13] The choice of a rural residence, at least amongst those who can afford it, suggests that the amenities the countryside offers are preferable to those of town and city. Unfortunately, many factors contributing to the amenity of the countryside have been destroyed or threatened with destruction by modern agricultural practices.

There may be hope for the industry if the product mix of farming is redefined by policy makers and farmers to include those elements that have been suppressed by the policy mechanisms and largely ignored by farmers in the recent past. This redefinition of the product mix must pay much greater attention to the changing needs of the population, with larger amounts of leisure time, higher incomes for alternative products and an interest in countryside that shows no signs of diminishing.

The demand for a rural setting for tourist activity is not new. What is new is the scale of private affluence and mobility that has enabled the public to reach the country. The demand for countryside tourism is not limited to the coastline or the rugged hills. There is also a demand for tourist facilities in the lowlands where mixed farming and deciduous trees survive to create a pleasant contrast to the arable dominated areas of other parts of the lowlands. The richness of the rural landscape as a tourist resource has been reduced by agricultural change and other forms of development. Yet it is tourism which is seen as an industry with great potential for growth. Unless care is taken of the countryside and the landscape is treated as a valuable product, its tourist potential will diminish rather than increase with time. Farmers are becoming more aware of their ability to transform the landscape. Where it is transformed, it should be done in such a way as to enhance its value rather than reduce it. The enhancement of landscape can yield benefits if farm tourist enterprises exist to reap the returns on what must be a collective investment by the farming community.

As is the case with tourism, the last 40 years have witnessed a remarkable expansion in countryside recreation, which again is attributable to rising levels of affluence and private mobility. Threatened by an invasion of urban hordes, a barricade mentality developed in the farming community. Instead of seeing the mobile millions as an opportunity, there was a tendency to hide behind barbed wire. This generalisation is unfair to those who recognised the potential for PYO enterprises and other attractions. Weaned on a diet of *Shell Guides* and *Reader's Digest* countryside manuals, and bolstered by Bellamy, Attenborough and others, these passive recreationists have become more concerned about, and interested in, the countryside they visit. The more adventurous have tried to venture into the countryside. All have seen a tendency for the recreational potential of the farmed countryside to decline. Farmers, stimulated by irrational policies, have been destroying a product that the public wanted, in order to produce more agricultural surpluses that are unwanted. In a mature, maybe moribund, economy the demands on rural land have changed and a wider range of products is sought from the countryside than is being offered by a supported and distorted farming industry.

The most heated debate about farming and the rural environment has been that relating to farming and wildlife. The debate culminated in the Wildlife and Countryside Act 1981, which has provided an uneasy truce. The passage of the Bill was marked by an unprecedented amount of public debate. During this time the amenity lobby acquired new skills of parliamentary lobbying and in the public presentation of its case. Hitherto, the farming community had proved a far more effective lobbying agency. However, a combination of tweeded traditionalists and urban activists joined forces to plead for wetlands and downlands, for heathlands and deciduous woods. Their pleas were listened to inside and outside parliament by growing numbers. Although nature conservation values are a product of an affluent society and are frequently held by the non-entrepreneurial middle classes, they have acquired new importance. They cannot be dismissed as an irrelevant luxury.

The interest in traditional countryside pursuits like riding, shooting and fishing has grown significantly in recent decades. At least some of the urban evacuees have acquired a taste for the traditional countryside recreational activities. At the same time, others campaign against what they see as barbarous activities masquerading as sport. Particularly in the case of shooting, and to a lesser extent with fishing, it has proved difficult to commercialise these gentlemanly pursuits. However, the proliferation of riding centres indicates a greater willingness to respond to these demands than those relating to wildlife or landscape conservation. In part, this is attributable to the ability to commercialise riding, but it also can be explained by a willingness of the farming community to dabble with the familiar, rather than recognise new and unfamiliar demands.

The interest in countryside extends further than an interest in recreation, tourism and conservation. In spite of the expansion of the supermarket in the food retailing sector, there has also been a marked increase in selling from the farm. The search for authentic products – either raw as in PYO products or with value added by processing – has led some consumers into the countryside to seek an antidote to the experiences of the supermarket. Trends in tastes for country products must be monitored and changes in retailing behaviour understood. Sales from the farm are likely to have an important niche in the future which at least a proportion of the farming community should be able to take advantage of.

Farm production has tended to focus, inevitably, on those products which are supported by the policies of the CAP. Consequently, products which are, or might be, in demand are shunned in favour of those which are supported. There has been an upsurge of interest in alternative crops[14] but it is characteristically a response to a policy-induced crisis, rather than a market-oriented search for viable new products. There are signs of growing sophistication amongst consumers, including interests in new foods and foods uncontaminated by chemicals or additives. These interests merit the attention of the farming community but the exploration of the possibilities must be underwritten by marketing principles rather than solely directed by technical considerations.

It may be possible for farmers to develop on-farm resources which have no current agricultural value. Derelict woodland or wetland areas may be developed into productive non-agricultural enterprises. In the future there may be a case for developing the forestry enterprise at the expense of farming. Afforestation is more likely to result from government policy than market forces. It should be recognised that timber is one of the few raw materials which is not in surplus in the EC. The absence of a forest tradition in the farming community is not sufficient justification for ignoring the possibilities of afforestation. Public endorsement, especially of broadleaved woodland, may be a partial substitute for market research but forest policy is bound to remain subject to governmental whims.

Farm buildings with minimal agricultural earning power may have alternative uses outside agriculture as industrial workshops or for conversion to tourist or permanent residential accommodation. Although there may be a demand for new workspace in some locations, any conversions should be preceded by thorough market research.

It would be remiss not to recognise the response of individual farmers and certain organisations to the new challenge. The recognition that consumers do matter and the realisation of new opportunities has been pioneered by a minority of farmers, often with minimal assistance from the advisory service. Furthermore, organisations like the Farming and Wildlife Advisory Groups have dabbled with landscape management and conservation issues, but their appeal to date has been primarily to the converted. The more recent words of the Country Landowners' Association and the National Farmers' Union have constituted an attempt to alter the corporate image of the farming community, but these words have often fallen on deaf or antagonistic ears. Nevertheless, the composite message cannot be ignored. A greater market orientation must permeate the farming industry. Painful adjustments will be necessary; adjustments in the way people think as well as the types of agricultural activity pursued. New enterprises will find a more important role in agriculture in the 1990s: enterprises that are more closely attuned to the desires of the consumer and enterprises that recognise the need for broader product mix than that created by the past rigidities of agricultural policies and the resultant actions of the farmers.

One problem policy makers must continue to address is the difficulty of making certain alternative goods and services marketable commodities. Clean air, pretty landscapes and interesting wildlife are all examples of what economists term 'public goods'. The private sector does not provide these goods or services at all effectively and this creates a demand for policy to create mechanisms to ensure that public demands are satisfied. The ESA payments or management agreements in national parks are examples of how farmers can be paid for providing these public goods. This is effectively a form of alternative enterprise where food and conservation products are produced together. Policy changes are beginning to focus more attention on public goods and alternative enterprises, but it remains to be seen whether the rhetoric of a market oriented diversified agriculture can be matched by adequate advice and action on the ground.

To summarise, the farming industry has been nurtured by an

agricultural policy that has allowed it to neglect consumers' wishes and adopt a production oriented approach with a narrow range of supported products. Rural Britain consists of more than an agricultural resource, and the need to look beyond narrowly defined agriculture is fundamental to the restructuring of rural economies from Australasia to North America and Europe. It is a resource with potential for tourism, recreation, conservation and silviculture that can provide living and amenity space for an affluent population with time and money on its hands. As those who own and work the bulk of Britain's rural land, farmers are in a unique position to respond to the new demands. Instead, a more intensive, more specialised agricultural sector has developed, reducing the availability of alternative products at the very time that they are being more sought after. Farmers, farmers' organisations and the Agricultural Development and Advisory Service must give more attention to the potential of alternative enterprises.

REFERENCES

1. BARKER, J. W. (1981) *Agricultural Marketing*. Oxford University Press.
2. BASTER, A. S. J. (1935) in: BAKER, M. J. (1979) *Marketing: an introductory text* (3rd edn). Macmillan.
3. BOWLER, I. R. (1980) *Direct Marketing in British Agriculture*. University of Leicester.
4. BUCHANAN, C. (1976) *A Tourism Strategy for the West Country*. Industrial Market Research Ltd.
5. See BAKER, M. J. (1979) *Marketing: an introductory text* (3rd edn). Macmillan, for a fuller description of psychological influences.
6. KOTLER, P. (1980) *Marketing Management* (4th edn). Prentice-Hall.
7. *FARMERS WEEKLY* (1983) 2 December.
8. MOOK, R. (1986) 'Identifying market opportunities using market window analysis', *Shii-take News*, 3, 2, 5–8.
9. These concepts are explained in Chapter 5.
10. RIMES, R. (1984) *Farm Holiday Accommodation in Devon: some aspects of marketing* (unpublished report).
11. BAKER, M. J. (1979) op. cit., p. 200.
12. TAYLOR, J. (1985) quoted in *Farming News*, 12 April.
13. HEDGER, M. (1981) 'Reassessment in mid Wales', *Town and Country Planning*, 50, 10.
14. ROYAL AGRICULTURAL SOCIETY OF ENGLAND/AGRICULTURAL

DEVELOPMENT AND ADVISORY SERVICE (1985) *A Guide to Alternative Crops.* RASE/ADAS.

Chapter 4

WHAT ARE THE ALTERNATIVES?

There is a bewildering array of possible courses of action confronting the farmer who wishes to diversify his business activities. But, as can be seen in Table 4.1, there are four main types of response that can be identified. It is possible to diversify on the farm or off the farm. The on-farm diversification can take the form of new types of enterprise which generate an income stream (if they are successful) or can consist of realising the value of assets which have a higher value in non-farming use than in farming. The acquisition of planning permission on barns is the obvious example of this. Usually the development will not be carried out by the farmer but the sale will have realised assets which might be used to ease indebtedness or to invest in diversification either on or off the farm.

Table 4.1 Types of diversification

	On farm examples	*Off farm examples*
Income stream	Snails (see Table 4.2)	Agricultural contracting, unrelated business activity
Asset realisation	Building land sales, barns for conversion	Selling shares

Off-farm diversification or *pluriactivité* in the language of the Eurocrat and Euroacademic is known to be a significant source of income for farm households. The multiple job holding that occurs in many farm households may or may not be related to farming. Agricultural contracting may be linked intimately to the farm with a common machinery pool providing for both businesses but in many cases the other gainful activities are likely to be completely unrelated to farming. It is important that farm households should recognise these off-farm options rather than lock themselves into unworkable on-farm diversification strategies. It is equally important that those guiding the adjustment process by policy instruments should not neglect the potential for developing off-farm employment possibilities.

Studies have revealed the very wide range of other gainful activities pursued by farm households. Ruth Gasson's pioneering work[1] has been followed by other studies which reveal the importance of off-farm earnings[2] and on-farm diversification to the welfare of farm families. A survey of Scottish farms indicated that over 40 per cent of farms had diversified business activities. Some 28 per cent of farm households had off-farm incomes and 19 per cent had diversified on the farm.[3] This figure of around 40 per cent of farmers ties in with estimates made by Produce Studies Ltd in 1988, although their studies are not directly comparable.[4] If anything, there is likely to be a rather greater percentage of diversified farm households in England and Wales.

Part-time farming has been looked down upon until recently by policy makers, and part-time farmers have been excluded from many agricultural support mechanisms. In spite of this they are growing in number, partly because of income pressures on farming and partly for personal reasons.

However, this chapter focuses on on-farm diversification possibilities in view of the almost insurmountable difficulties of generalising about the non-farm skills of farm households and the potential applications of these off the farm. No farm household should neglect to examine the off-farm options. But alongside this review a thorough appraisal of the on-farm options is called for.

The basis for an analysis of these alternatives must be a recognition of the demand prospects. With certain alternatives, location is likely to be a crucial influence on demand.

The focus on demand should not lead to a neglect of production problems. It is not always reasonable to assume that farmers have the required skills to be able to establish and manage alternatives. Nor can it be assumed that adequate technical knowledge of production systems exists. The lack of technical information for many alternative enterprises is highlighted in the recent Centre for Agricultural Strategy study of alternative enterprises.[5] This study offers a rather more production oriented approach to the classification

Table 4.2 The main groups of alternative enterprises on farmland

Tourism and recreation	Tourism	Bed and breakfast Cottages/chalets Caravans/camping Activity holidays
	Recreation enterprises	Farm museums Visitor centres Riding Game shooting Other shooting Fishing Farmhouse catering
Adding value to conventional farm products	Animal products	Meat (direct sales, etc.) Skins/hides/wool Dairy products (direct sales/processing)
	Crop products	Milled cereals PYO and direct sales of vegetables
Unconventional agricultural enterprises	Animal products	Sheep milk Rare breeds Fish Deer and goats
	Crop products	Linseed Evening primrose Teasels
	Organic production	
Use of ancillary buildings and resources	Woodland products	Fuel wood Craft timber products
	Redundant buildings	Industrial premises Accommodation
	Wetland	Fish Game
Public goods	Wildlife	ESA payments
	Landscape	Management agreements
	Historic sites	Heritage relief
	Access	Access agreements

used subsequently but provides a useful introduction to the full range of alternatives. Fish farming is not just another form of animal production, and crops like evening primrose and borage have presented the early producers with real production problems. However, in the absence of existing or projected demand there is no justification whatever for contemplating alternatives. Once a demand has been recognised the production problems can be rooted out.

The main groups of possibilities are summarised in Table 4.2. The table offers a grouping of alternatives, not a comprehensive list, and examples from each group will be examined in the text. Each group will then be examined in turn. The reader who is not interested in the alternatives within each group may ignore the subsections.

TOURISM AND RECREATION

There is a long tradition of the use of the countryside for tourism and recreation. Rented accommodation was available in upland Britain in the nineteenth century and for a far longer period farmland has provided a recreational opportunity for such activities as hunting and shooting. Until relatively recently, the numbers involved were small. More recently, with greater private mobility and additional leisure time, the use of the countryside for leisure and recreation has increased. Farmers have developed ambivalent attitudes to this increase, sometimes exploiting the market opportunities created but sometimes acquiring a siege mentality and an antagonism towards visitors.

A major reason for the emergence of these ambivalent attitudes is the fact that many of the benefits derived by tourists and recreationists are external to the market economy, i.e. no money transactions take place. Benefits from viewing the countryside or from walking the footpath system are external benefits. At the same time as acquiring the benefits, the recreationists may impose certain costs on the recipient areas and it is this factor that has tended to arouse the resentment of rural communities. The challenge to farmers, and rural communities in general, is to 'internalise the externalities', to produce goods and services which recreationists are prepared to pay for. This is relatively easy in the case of tourism but presents more difficulties in the case of recreation.

TOURISM

As an area of economic activity, tourism has often been stigmatised by both farmers and the wider community. The reasons for the stigma are complex and can be attributed partly to facts and partly to attitudes. Tourist activity has always been seasonal and changes in taste, shifting exchange rates and an unreliable climate combine to produce a degree of uncertainty about tourist prospects. These facts are irrefutable. However, attitudes to tourism, especially in the farming community, seem to be more negative than the facts would merit. True farmers are seen as tillers of the fields, not tenders of the tourist flocks. Where tourism is a farm enterprise it is frequently seen as the domain of the farmer's wife. A philosophy of agricultural fundamentalism influences both attitudes and actions. If farmers were rational, economic men, their interest in tourism might be greater.

In spite of the uncertainty that surrounds tourism, it has been an industry that has fared relatively well in the turbulent economic climate of the last decade. Spending by domestic holiday makers in the UK increased from £3100 million in 1978 to £6727 million in 1987. Over the same period, overseas tourists' spending in the UK increased from £2479 million to £6273 million.[6] However, the tourist industry contains many component parts. The prospects and problems of different sectors and different regions are likely to vary considerably. It would be misleading to make assertions about rural tourism in general, and farm tourism in particular, on the basis of generalised figures of tourist trends for the entire country. Nevertheless, the frustration of airline travel, safety fears and tourist 'blight' in certain parts of the world augur well for domestic holidaymaking.

Although, in money terms, expenditure by tourists in the UK has increased continuously since the early 1970s, this is not the case if inflation is taken into account. In the late 1970s and early 1980s there was a decline in tourist activity which affected countryside areas as well as traditional resorts. This trough in spending bottomed out in 1981/82, since when spending has increased. Overseas visitors' spending has increased from £3500 million (1984 prices) in 1982 to £5490 million in 1987, whilst the increase with domestic tourists is less marked with a growth in spending from £3950 million in 1982 to £4850 million in 1987. At a national level, there was no evidence of countryside districts increasing their share of tourist activity, but in Devon, the decline was most pronounced in the resorts, and tourism remained relatively buoyant in the countryside. Since 1980, rural areas of Devon have increased their share of the county's bedspaces from 17 to 26 per cent. This is likely to have reflected changing consumer preferences. At times of decline, some rural districts are likely to suffer. Remoter areas are most likely to lose out and the problems of remoter areas are often more acute when fuel prices rise abruptly.

The key to the continued expansion of tourism in the UK and in many other parts of the world is held by many to be 'heritage'. The imagery is evident in UK government reports on tourism, where thatched cottages and village pubs vie with castles and heritage centres for the most photographed attractions in the latest glossy publication. In the US, reconstructed colonial settlements provide major tourist attractions. Where regions rich in heritage coincide with areas rich in natural endowments of scenery or wildlife, the package is complete. All that is needed is a structure of accommodation and ancillary facilities to satisfy the public's apparently insatiable appetite. Rural areas and farming communities have opportunities to tap into this vein of heritage interest but it must be recognised that the providers of major facilities may be located in urban, rather than rural, areas.

One feature of tourism in the UK in recent years has been the development of tourist ventures, often with substantial public sector backing, in places hitherto regarded as unlikely tourist destinations. Urban centres like Bradford, Manchester and Glasgow have appeared on the tourist map. There is also evidence that the dominance of London in terms of overseas tourists is declining. These trends should offer food for thought for potential tourist businesses.

In general, but not without certain exceptions, the prospects for rural tourism might be seen to be more favourable than for tourism as a whole. The sector of the industry that has experienced the greatest decline has been the traditional seaside resort, particularly those with rural hinterlands with few attractions. The quantity of serviced accommodation in these resorts has declined and there has been a significant trend towards self-catering in the resorts and throughout the country. In the early 1980s the consensus of opinion was that there was a continued demand for self-catering accommodation.[7,8] By the late 1980s, there were increasing doubts and a study of rural tourism in 1987 commented on 'the knife edge on which many of these businesses existed'.[9] This observation merely reinforces the need to be cautious about generalising about tourist prospects. However, a general 'greening' of public opinion is likely to be associated with an interest in countryside. If the right type of countryside can be marketed, the prospects for rural tourism must be reasonable.

Tourist operators in rural areas stand to benefit from a number of changes in the patterns of holidaymaking. The growing demand for self-catering accommodation has been noted. This could be met by building conversions or improvements and the provision of purpose-built chalets. There is also a growing interest in activity holidays, particularly for young people. The countryside provides the setting for a wide range of traditional and new activities. A third trend of significance is the growing number of second or third holidays and short-break holidays. Rural districts reasonably accessible from major urban areas are well placed to cater for this demand.

The farmed countryside provides the setting for rural tourism and for day visits from tourists in urban areas, who explore the countryside from an urban base. Their interests in an attractive countryside with a rich wildlife may be somewhat at variance with those of the farmer. A preoccupation with production is likely to generate a landscape that tourists find less attractive and the tendency towards specialisation in agricultural production may reduce the variety in the landscape still more. By eliminating hedgerows and copses, farmers must be reducing the scenic potential which may be converted into income through the development of tourist enterprises.

At a general level the demand for rural tourism looks likely to be sustained in the immediate future. Rural Britain will remain a destination for holidaymakers as a place to stay and a place to visit. However, blind optimism is not justified. Exchange rates can alter, making overseas tourist destinations more economic, and overseas farmers, too, can offer rural holidays. The highly developed gîte system in France is a potential competitor. The French Ministry of Agriculture has funded and supported the development of farmhouse holidays for 30 years. By 1979 there were 28,000 gîtes, many of which had adapted to market demands by providing for particular clienteles.[10]

The support of tourist ventures from a ministry with a specifically rural remit may be a distinct advantage. Tourist planners tend to think in terms of large-scale projects but there may be smaller scale alternatives kinder to the environment and more beneficial to the rural population. The Swiss have pioneered what is termed 'soft' or 'green' tourism. This unobtrusive, low environmental impact tourism can be organised at the grass roots, rather than imposed on a district. It requires no less a professional approach than the

Plate 4.1   *Serviced accommodation in farmhouses*
Many farmhouses are large. Unutilised rooms form a potential
resource that can be adapted for tourist use with minimal capital.

Plate 4.2   *Self-catering accommodation*
The decline in the farm workforce has left many farm cottages
unoccupied. Holiday letting may provide a more flexible option
than selling.

alternative 'hard' tourism projects, but it is likely to appeal to the discerning visitor who wishes to distance himself from the holiday hordes.

One sector that has exhibited considerable growth in the last 15 years has been touring caravans and tents. Farmland provides a potential camp site and many farmers have developed camp sites for touring caravanners and campers. In Devon, the number of touring caravans at peak season increased from around 3000 in 1970 to nearly 10,000 in 1984, since when the numbers have been more volatile, principally because of the climate. Consumer preferences for this type of holiday are a function of its cheapness, its flexibility and the independence it offers to the holidaymaker. Planners, amenity groups and other motorists often regard touring caravanners less than favourably. In spite of the fall off in numbers, there may still be potential in some areas for additional facilities. The establishment of small-scale, short-term, camping and caravan sites for periods of up to 28 days or under the auspices of the Caravan Club can avoid any planning permission requirement. This might allow interested persons to test the water.

Few surveys have attempted to establish with any precision the market segments formed by class opting for holidays on farms, and their motivations for visiting farms. Frater's evidence from survey work in the English Midlands points to affluent middle- and upper-class families as the largest group.[11] There are likely to be differences between self-catering and serviced accommodation in that families with older children may be more likely to use the latter. Most of those who visit farms have stayed on them before and are committed to the idea of a farm holiday before they acquire the brochures. They choose to stay on a farm not because of farming but because farms are in the countryside, in the peace and quiet and offer good value for money (see Table 4.3).

Table 4.3 The attractions of a farm holiday

Attractions	(%)
Country life	27
Peace and quiet	24
Value for money	23
Children's experience of farm life, etc.	10
New experience	5
Farm activities	4
Others	7

Source: Frater[11]

The findings in Table 4.3 were substantiated by survey work in Devon carried out by Rimes.[12] In the Devon survey some evidence was collected which indicated that farmers' and farmers' wives' expectations of what holidaymakers wanted actually differed from what the holidaymakers felt they were seeking. These observations raise two important questions. Firstly, can more people be stimulated into taking a farm holiday by improved promotion? Secondly, do the misconceptions of farm tourists' interests held by farm tourist operators adversely affect either the visitors' experience or the way in which the farm holiday is promoted? If visitors' needs and interests are catered for and the product mix adjusted accordingly, farm tourist operators will be responding to one of the most important marketing principles: an understanding of the needs and demands of the consumer.

The supply of farm tourist enterprises is likely to be influenced by many factors which include social, economic, legal and planning factors. Negative attitudes to tourism and townees are likely to be major reasons for lack of involvement in tourist enterprises. Even without negative attitudes many families may be unwilling to accept the level of intrusion created by paying guests in serviced accommodation. On the other hand, a positive desire to mix and meet people may lead to the provision of farm tourist accommodation for non-economic reasons.

The economic reasons for creating farm tourist facilities have become more prominent in recent years and are likely to remain important. As farm incomes come under additional pressure, so there will be a need to respond to any opportunity that presents itself. Unconventional enterprises are likely to assume a position of greater importance. The economic pressures that affect the supply of farm tourism must be balanced against the economic laws that will control that supply by eliminating the unprofitable enterprises.

The single greatest threat to small-scale farm tourism is likely to be oversupply rather than a change in demand. Optimistic forecasts of the demand for self-catering accommodation and the additional 'push factor' of declining incomes from conventional farming may have created the potential for oversupply. The diversification grants offer a further push to the supply curve. However, talk of oversupply must be tempered by evidence that, in the right place and with the right quality of accommodation, highly successful enterprises are prospering. Those enterprises that obtain low levels of repeat bookings or personal recommendations should be asking themselves serious questions about the product they offer.

The provision of self-catering accommodation for tourists is at times motivated by reasons having little to do with tourism. Various Housing Acts have created potential difficulties for landlords and the decline of the agricultural workforce has frequently left farmers with vacant property. Often property is improved not for the sole purpose of creating a tourist facility but to increase the capital value of the holding and provide a potential residence for members of the farm family at some future time.[13]

Even the motivated household which confidently predicts profit from the contemplated enterprise may be constrained by factors beyond its control. Clauses in tenants' leases may restrict such enterprises, although some landlords are beginning to think more flexibly. Planning permission is required for anything more than small-scale farmhouse accommodation and short-term or certificated caravan and camp sites. Table 4.4 summarises the planning and other legal requirements that must be countenanced with a farm tourist enterprise. Planning authorities' attitudes to rural tour-

Table 4.4 Legal and planning requirements of tourist enterprises

Type of development	Legal and planning requirements
Farmhouse guests	May require planning permission if it involves much of house and residence is subsidiary
	For over six people or second floor bedrooms, fire certificate required
	Food hygiene regulations
	Business insurance if more than six guests
Self-catering accommodation	Planning permission required
	Building regulations for any conversion work
	Fire certificate (as above)
Camping and caravan sites	Planning permission required except: three vans for twenty-eight days any number of tents for twenty-eight days caravan clubs/camping clubs certificates of exemption
	Licence from local authority – covers numbers, location, amenity, fire, sanitation

Sources: ADAS socio-economic advisory leaflets
English Tourist Board development guides

ism range from active promotion, through benign neglect to stubborn opposition. In spite of the growth in touring caravanning, it is normally viewed by planners as a threat to amenity, while self-catering units converted from barns may be more favourably received. Reluctant as the farm household may be to get involved with planners, it is necessary. Early consultation can eliminate problems. It is easier to make minor adjustments to business strategy at the outset than after being served with an enforcement notice once established.

Support is available in the form of advice, the possibility of grants and loans, and in occasional short courses relating to tourism. In contrast to the provision of agricultural advice, grants and training, assistance relating to tourism is limited. Few agricultural advisers are particularly interested in, or knowledgeable about, tourism and tourist board advisory services can appear somewhat remote from the small operator. The grant aid or loans are likely to be for larger projects and the whole tourist industry competes for slices of a small cake. Grants for tourist accommodation have been available from various sources at various times. The Farm Diversification Scheme, Rural Development Commission grants in Rural Development Areas and, until recently, 'Section 4' grants from the English Tourist Board, were all available. The combined effect of these grants on the supply curve for farm tourism is unknown. The highest levels of grant are available from MAFF but the low ceiling means that, for many larger tourist projects, alternative sources of grant would be preferred.

Education and training for tourism in general has not been especially effective. The House of Commons Trade and Industry Committee commented in 1985[14] that:

> the provisions for training for a career in tourism are pathetic. . . . the tourism industry requires large numbers of people, few of whom need special skills, but nearly all of whom require some basic training.

Efforts have been made to remedy this with short courses in local technical colleges and an expansion of full-time provision, but there remain many deficiencies needing to be tackled. But many agricultural colleges and the Agricultural Training Board have begun to address this need. The success of courses should be measured by their effectiveness in improving the efficiency of the enterprises of those who attend them. No such evaluation has been conducted to date.

The present pattern of involvement of farmers and their wives

Plate 4.3   *New forms of self-catering accommodation*
This 'rest house' has been constructed near the start of the
Southern Uplands Way at Abbey St Bathans, Berwickshire. It caters
for long-distance walkers or car-borne tourists and offers a very
high standard of accommodation.

with farm tourism is unknown. In 1974 a study by E. T. Davies
estimated that between 4–6 per cent of farmers in England and
Wales were involved in tourist and recreational enterprises.[15] This
amounts to between 10,000 and 15,000 farmers. Evidence from
Less Favoured Areas for 1981 indicates a much higher level of
involvement at about 20 per cent of farmers in these areas.[16] These
figures of overall levels of participation mask very major variations
in the scale of enterprise from the insignificant to the large.
Undoubtedly, the total contribution to UK farmers' incomes from
tourism is considerable. The total receipts from farm tourism in the
UK were estimated in 1974 at £40–50 million, though this includes
recreation. This figure seems high if Davies' farms in Less Favoured
Areas are typical tourist farms over the country as a whole.[16] At
1981 prices, £50 million would be the upper limit of an estimate
for tourist income alone and, in real terms, this would represent a

drop from 1974, which seems unlikely, although Davies' figures relate to only one particular year.

Several attempts have been made to estimate the profitability of farm tourist enterprises. The most detailed work has been carried out by Davies at Exeter University. This culminated in the study of farm tourism in Less Favoured Areas in England and Wales. Because it covers a wide area, it will be used as the basis for the following discussion. Earlier data, such as that collected by DART for their 1974 study, are of historical interest only.

Data were collected on capital tied up in tourist enterprises and the return on this capital (see Table 4.5). A very wide range of average returns is found, ranging from 8 per cent on self-catering conversions to 141 per cent on small caravan and camping sites. Ironically, the return on capital is least in the very enterprise which farmers have been most strongly advised to enter. This may reflect reasons for building conversions, other than just profiting from tourism, amongst those who have already developed such enterprises. In addition, the amount of capital tied up in self-catering units is high compared with the capital requirements of a bed and breakfast enterprise. However, the absence of any negative returns and high average returns for many enterprises is a healthy sign.

Enterprise types can also be compared by examining their margins over direct costs. This figure is broadly similar to a gross margin but, in view of the relative ease of allocating costs, the 'direct costs' include certain items such as maintenance, rates and fuel which are usually categorised as fixed costs. The direct costs are, as their name suggests, those costs which can be directly allocated to the enterprise. For comparative purposes, it is easier to examine the margin per £100 receipts (see Table 4.6). The lower levels of margin cannot be taken to indicate low profitability but may merely reflect the higher level of allocatable costs of such items as food. Serviced accommodation requires a low capital investment but high running costs, whilst self-catering accommodation has high capital and low running costs.

The most interesting data provided by Davies show the range of returns to the same enterprise type to be remarkably variable (see Figure 4.1). Unfortunately, one can only speculate on the causes of these variations. The variability is likely to be a function of factors within the operator's control (the product mix) and of factors that cannot be controlled. The cost of the meals can be varied; the location of the farm must normally be taken as fixed. Anyone contemplating or managing a tourist enterprise should be aware of this wide range in profitability and endeavour to identify the factors that can be controlled to influence profitability. Where profitability is not the main reason for involvement in farm tourism, a more tolerant attitude to low returns may be justified. At the same time, it may be desirable to be aware of the costs of the enterprise, particularly if it is soaking up scarce resources on the farm.

Table 4.5 Annual return on capital of farm tourist enterprises in Less Favoured Areas

	Annual return on capital (%)
Bed and breakfast } Bed, breakfast and evening meal }	37
Conversion for self-catering	8
Static caravans	12
Touring caravan/camp sites: small large	141 13

Source: Davies[16]

Table 4.6 Margin over direct costs per £100 receipts in Less Favoured Areas

	Margin per £100 receipts (£)	Range margin per £100 receipts (£)
Bed and breakfast	48.1	
Bed, breakfast and evening meal	51.1	20–76
Conversion for self-catering	75.6	49–91
Static caravans	67.5	40–89
Touring caravan/camp sites: small large	94.7 79.0	Not available 44–86

Source: Davies[16]

Farm tourist enterprises have been advocated as a means of boosting incomes of farmers who, for a variety of reasons, experience low levels of income. Where the reason is a lack of business skills, it is unlikely that a shift to an enterprise which demands a strong market orientation will provide the road to salvation. Where

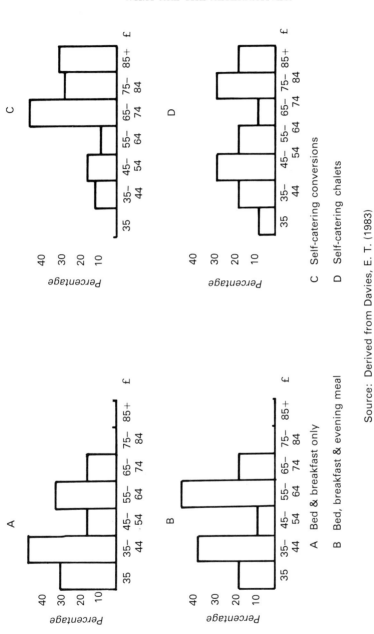

Source: Derived from Davies, E. T. (1983)

Figure 4.1 The variability of margins per £100 receipts in selected tourist enterprises.

the reason is a lack of agricultural resources and an abundance of tourist resources, then farm tourism can be seen as a useful development tool. Some evidence suggests that it is not those most in need of additional income that enter farm tourism. The Less Favoured Areas farmers participating in tourism run larger farms than the average for Less Favoured Areas.[17] However, in a Herefordshire study, the average farm size of farms participating in tourist enterprises was below the average for the area. It is difficult to argue a case for farm tourism as a small farm development strategy, for the benefits of existing resources in spare buildings, farmhouse space and available capital are likely to remain with the larger farmer. Furthermore, larger enterprises stand more chance of getting tourist board grants and advisory support. It would need a policy of positive discrimination in favour of small farmers to make farm tourism a major enterprise for this group. This assertion in no way dismisses the possibility of tourism as an enterprise in smaller farms. It would be unfair, however, not to put the opportunities into perspective.

Farm tourism can create income earning possibilities. The rewards can be considerable. If the full benefits of farm tourism are to be derived, farmers and their advisers must shift much more closely towards the marketing approach. There are encouraging signs. The emergence of a Farm Holidays Bureau is a welcome step. So, too, have been the successful attempts to develop farm tourism in areas that do not immediately figure in the tourist's mind. The work of various official groups (including ADAS) in north Staffordshire is evidence of what can be achieved in an environment not normally associated with tourism. Alongside these positive signs, old negative attitudes linger. Tourism is viewed with suspicion: the production orientation still prevails. As the fragility of the production orientation is exposed by contemporary events, so tourism should be explored more rationally and from a marketing perspective by the farming community.

The expansion of interest in farm tourism may create new problems. Established farm tourist enterprises may find their profits and welfare threatened by new entrants who may be more finely attuned to changing demands. The Farm Holidays Bureau and its member groups may find the optimism of youth replaced by the bickering of adolescence. The relationship of local groups and the co-ordinating body may create problems, in that local groups may not see a direct return for their contribution. Furthermore, within local groups there may be significant variations in the quality of accommodation and, as yet, no grading system has been devised to

Plate 4.4   *Marketing farm tourism*
            The Farm Holiday Bureau acts as a central agency for promoting
            farm tourism. As with any central marketing agency, it is vital that
            individual members can see the benefits that the agency produces.

help the holidaymaker choose the right place for his holiday. Local
groups cannot afford to carry 'passengers' whose standards of service
and accommodation do not reach desired levels, especially in the
absence of an externally vetted grading system based on random
inspections.

The future of farm tourism is assured but the survival and development of individual tourist enterprises will depend on a number
of factors. A good location is a major advantage but this cannot
make up for deficiencies in hospitality and financial management
which must underpin successful enterprises. As competition stiffens, so the weaker enterprises will be thinned out. And, in the
likely environment of more competition, decisions must be made
as to whether to go it alone and market an individual enterprise or
market farm tourism through one of the many marketing groups.
The adept operator may be able to stitch up deals with travel
companies, whilst the more passive provider might accept the package provided by a farm holiday group. Optimism about the scope
for tourist enterprises must be tempered by a realistic attitude
towards the managerial demands.

RECREATION

Whilst tourists have always paid for their services, recreationists in the countryside frequently have not. The fact that so much country-side recreation has a weakly developed market dimension has tended to lead to its neglect by farmers. Furthermore, the fact that outdoor recreation opportunities are heavily influenced by the weather and that use is subject to marked seasonal peaks, has generated further negative thoughts. In spite of these negative factors, the outlook is by no means bleak. Large numbers of people visit the countryside and have shown themselves willing to pay for certain types of recreation. There are also certain responsibilities for landowners and farmers relating to the non-paying types of recreation. They are the guardians of the landscape and of the footpath system. If that guardianship is not exercised with due regard to public attitudes, the image of the farming community can be tarnished. This has undoubtedly been the case.

The demand for different types of recreation must be recognised and the extent to which this demand can be satisfied by on-farm facilities should be considered. A wide range of opportunities exists (see Table 4.7). Overall, there has been a marked increase in the demand for countryside recreation in the last two decades but the pattern of expansion has been different for different types of activity. The possibilities are inevitably constrained by planning and other regulations (see Table 4.8).

Many a townsman, suitably clad in Barbour and green welling-tons, has turned to traditional countryside recreation activities like shooting and fishing. Horse riding, too, has witnessed a remarkable rise in popularity. The conversion of urban fringe acres to 'horsey-culture' is evidence of this replacement of the work horse by the pleasure horse. In the late 1970s, there was a decline in some of the passive recreational pursuits but the traditional countryside sector has shown a growth in demand both as an activity and as a spectator sport. Such activities as sheep dog trials attract public interest on the television screen and in the field. There is nothing surprising in this interest in countryside sports. Thorstein Veblen[18] argued at the end of the last century that increased affluence would create a society of leisure and that leisure patterns of the masses tomorrow would be an attempt to emulate those of the squirearchy today. The interest in these traditional pursuits is frequently depen-dent on reasonably high incomes. The growth in traditional pursuits will only be halted if the recession reduces the real incomes of everyone. To date, what has tended to happen is a growing disparity between those with wealth and those without, the implications of which continue to challenge leisure forecasters.

Although demand for traditional countryside pursuits has grown rapidly, the transition from non-market activities has not been easy. In many cases this is likely to be a function of the muddling of business objectives on the part of the person running the enterprise. The suggestion that some of those who run riding centres do so for love rather than money is borne out by survey findings.[19] The same is likely to be the case where farmers run shoots or fishing enterprises. The available evidence on returns to sporting enterprises shows a predictably high variability but includes examples of highly profitable enterprises.

Where a businesslike approach is adopted to the development of such enterprises they can be profitable, especially where they can take advantage of areas of land with limited agricultural value. Undrainable wet areas can be turned into fishing lakes which generate an income far in excess of their alternative earning power in agriculture. Riding centres in the urban fringe or the uplands may generate returns greater than farming, even on better land in these

Table 4.7 Possible recreational enterprises for farms

Traditional countryside pursuits	*Active*	Fishing
		Shooting
		Hunting
	Passive	Museums
		static
		working
		Viewing/spectating
		Traditional events
Farming as recreation		Farm open days
		Farm trails
		Farm visitor centres
		Pick-your-own
		enterprises
Catering facilities		Farmhouse teas
		Restaurants
New activities on farmland		Grass ski-ing
		Hang-gliding
		Motor cycle
		scrambling
		Pop festivals
		Adventure games
Educational use of farmland		School visits
		Study centres

Table 4.8 Planning and other legislative requirements for recreational enterprises on farms

Traditional countryside pursuits	*Active*	Some activities require planning permission Water authority regulations control water body creation/stability, etc Riding centres must have local authority licence
	Passive	Planning permission required if activity is carried out for 28 days or more Health and hygiene on ancillary enterprises
Farming as recreation	*Open days*	No planning permission if less than 28 days per annum
	Visitor centres/museums	Planning permission required
Catering facilities		Planning permission required normally Health and hygiene regulations
New activities on farmland	*Grass skiing*	Planning permission required if operated for more than 28 days per annum
	Motor cycle scrambling	Maximum of 14 days per annum Otherwise, planning permission required
Educational use of farmland	*School visits*	Planning permission not required but care must be taken over safety
	Study centres	Planning permission required

Plate 4.5 *Riding as an alternative source of income.*
Riding enterprises may represent a more profitable enterprise than many upland farm enterprises. However, the highly variable profitability of riding centres should be borne in mind.

areas. The provision of riding facilities is likely to be increased by the farm diversification grants and the set-aside provisions. In the case of set-aside, the use of the headland set-aside provision may afford opportunities to create private bridleways – a good example of how externalities may be internalised.

Shooting enterprises on farmland range from the dramatically unprofitable to the occasionally remunerative. It is impossible to generalise about the possibilities for shooting. Much depends on the layout of the farm and areas of land suitable for shooting. Rough shooting may generate a small income for a negligible input but an intensive shoot is a substantial investment. Profitability of sporting enterprises with a substantial investment such as deer stalking, and grouse, pheasant or partridge shooting depends on sound management and astute marketing. There may be a sizeable overseas market that can be exploited but the right product mix must be found, and it is clearly not a market that can be tapped by the small farmer.

The interest in clay pigeon shooting has risen substantially in recent years. Location is crucial. Not only must the site be accessible but also it must be sufficiently remote from residential areas and public rights of way to avoid antagonising planners and public.

There is, in addition to a growing interest in active traditional countryside pursuits, a strong interest amongst the general public in passive activities relating to the traditional countryside. This is reflected on the television screens in numerous countryside programmes and in the countryside itself in museums of various types.

These museums vary greatly in their objectives, their size and their scope. The common binding thread is that they satisfy an interest in the past.

Acton Scott Working Farm Museum is a museum run by the local authority which is particularly oriented to the needs of an educational clientele. It is unlikely ever to make a profit, but this is no indictment for it provides an extremely valuable service. It also contains some of the ingredients likely to make a highly successful museum of any type: stock is accessible and visible, and there are normally activities such as butter making or working with shire horses taking place; additional services such as a cafeteria and gift shop are available and helpful and pleasant staff are on hand to help visitors.

Museums of rural life can too easily become a clutter of inanimate objects. Static displays might interest the experts but bore the average visitor. The latter can be catered for without unduly compromising the more conventional education bias of many museums. A unity of themes is a desirable quality and the explanation of displays should relate to the visitors whose interests must be captured and sustained. If working exhibits are not the norm at such museums, it may be possible to put on occasional displays of working machinery or animals or run theme days which will broaden the clientele. This approach will only work if supported by effective advertising.

It is impossible to generalise about the saturation point for farm museums. As more are established, so competition will intensify. Private sector operators are inevitably competing with local authority concerns where the ratepayer can cover any losses. A product mix must be offered that will attract visitors in the first place and succeed in making them return, and in making them stimulate visits from friends. Changes may be necessary as competitors raise standards and imitate successes. The farmyard collection of rural memorabilia, stacked haphazardly around some buildings, cannot be expected to be successful; a museum with a sense of purpose and a recognition of customer needs might be.

Modern farming can also attract the interest of the public. In the late 1970s and early 1980s farm open days were promoted widely as a means of educating the visitor about farming in an enjoyable way. These open days were grant aided by the Countryside Commission but research in Scotland suggested that visitors who attended such open days were already enlightened, whilst those in need of enlightenment pursued activities elsewhere. Those concerned about the image of the farming industry have revitalised the

Plate 4.6   *Interpreting the landscape*
A simple signboard clearly indicates the principles of crop rotation
to the visitor to Acton Scott Working Farm Museum. The landscape
offers all the elements desired by the visitor: mixed farming;
hedges, hedgerow trees; woodland. The stooks remind us that it
is an image of the past rather than the present.

open day idea for UK Food and Farming Year (1989). Countryside
Commission funding of farm open days has ceased, but the idea of
farm open days should not be dismissed and such events may, if
planned appropriately, attract different groups of people and offer
a range of services to the visitor. It should be possible to have open
days relating to a specific theme such as a sheep day at lambing

time. Hand spinning and weaving might offer interesting ancillary demonstrations and possible outlets for craft workers' products. The success of farm open days is highly dependent on the skills of the farmer in communicating in the right way to the public, and recognising different levels of interest and understanding in different individuals or groups. Palgowan, near Glentrool, entertains and involves the public and is a good example of a farm that is prepared to entertain as well as educate, and obtains a small income from so doing. There are many other necessary considerations including visitor safety, way-marking and advertising which are well summarised in a Countryside Commission manual.[20]

Farm visitor centres are permanently (or semi-permanently) open farms which have usually invested in substantial capital to accommodate visitors. Such farms as Dairyland in Cornwall or Easton Farm Park in Suffolk provide a wide range of visitor attractions, including modern farming practices, as well as more traditional attractions like museum exhibits. Lightwater Valley in Yorkshire was established to provide a commercial visitor enterprise which includes a serious attempt to explain modern intensive pig, poultry and beef production in a way attractive for the visitor. It is impossible to generalise about profitability of such centres. A well-managed enterprise in the right location can definitely be profitable but such enterprises normally demand major injections of capital and are not ideal for those that are either averse to risk-taking or short of capital. This type of development has become increasingly common. Often milk quotas provided the stimulus to push farmers into visitor enterprises. The Milky Way in north Devon or Oakwood in southwest Wales are recent entrants whose optimism in the expanding demand for leisure would appear to be justified. As the number of visitor attractions in holiday areas expands, so the essential qualities of entertainment and value for money will be exposed as competition stiffens.

The recreation element in PYO enterprises has already been discussed. It is most unlikely that it will become less important with time and the provision of a wide range of recreational opportunities may be a means of product differentiation at a time of increasing competition.

Farms can offer catering facilities to day visitors as well as tourists. In a county like Devon or Cornwall there are many farms offering traditional farmhouse teas which can provide alternative enterprises for the farm family. Location is likely to be very important if the principal market is the casual tourist trade. The extent of tourist and recreation traffic on rural roads is usually all too familiar

Plate 4.7   *Day visitor facilities on farms*
Dairyland in Cornwall is suitably located in a tourist region. In addition to offering the visitor the chance to view milking in a rotary parlour, the farm offers a variety of facilities. If children are happy, parents will often be satisfied. The junior JCBs are popular in the play area and are sold in the shop.

to the farmer. Tourist routes and recreation destinations can be recognised. Farms located on or near major footpaths – such as coastal footpaths or major inland walking routes – may be well placed to offer catering enterprises. If the customers are more local in their origins, then a good reputation may overcome a less advantageous location.

Catering enterprises on farms are likely to be of two types. Firstly, there are the small seasonal enterprises that can be run from the normal farm kitchen without undue difficulty. Secondly, there are more elaborate restaurants which are likely to require planning permission and will have to compete with the full range of public houses, hotels and restaurants. The attractions of a farm setting do allow a degree of product differentiation and the positive image of wholesome country food is also likely to help.

Farmland may provide a suitable venue for activities quite unrelated to farming. Minority activity sports such as hang-gliding, grass ski-ing or motorcycle scrambling require land. These various activities have their own particular requirements, such as close tight sward on suitable slopes for grass ski-ing, or suitable relief for launching hang-gliders. In the case of motorcycle scrambles, land of minimal agricultural value may provide highly appropriate tracks.

Adventure games have attracted widespread publicity as an alternative recreational use of land. They, too, take advantage of land of minimal agricultural value. However, the control of these is largely in the hands of franchisers, who have been adept at retaining a substantial proportion of the profits for themselves.

Numbers of people visiting the countryside have not increased inexorably over the last decade. The evidence points to a decline in countryside visits in the late 1970s followed by an increase in the early 1980s.[21] Since 1984, there has been a tendency for numbers to decline[22] but whether this is evidence of a trend or of bad summer weather is not clear. The latter explanation is more likely, but those contemplating day visitor enterprises should recognise the volatility of demand. Although the active countryside pursuits attract a relatively small clientele compared to passive activities, they appear to form a growing market. The effective development of enterprises of this type hinges around effective marketing and a flair for the recognition of non-farming opportunities on particular areas of land. The total involvement of the farming community will always remain small but, for those who have the appropriate facilities, possibilities may exist.

Farms and farm centres or museums can be used educationally. Catering for the educational market alone is unlikely to lead to maximum profits but there may be hidden benefits arising from better understanding of the countryside, and the image of farming may be promoted. It would be misleading to suggest that provision for educational groups is entirely unprofitable. School parties can comprise between 20–50 per cent of the total clientele of farm visitor centres, with the normal proportion being around 25 per cent. This is usually at non-peak times and helps to provide a more even flow of visitors. Satisfied schoolchildren might also entice their parents into taking them on a return visit. However, a major problem has arisen in the 1988 Education Reform Act. One provision of the Act makes it much more difficult for schools to charge pupils for visits. Many schools in the UK are extremely concerned about this 'rogue' component, which could lead, perversely, to the cancellation of many school visits which enrich the education of children. Whatever the outcome of the Act, teachers' needs must be met. Some centres produce schools packs (which are not always particularly well prepared from the teachers' point of view). Liaison before the visit is essential and it is preferable that schools packs should be an aid to teachers' preparation, not a substitute for it. A complete educational package should not be presented as a *fait accompli*. Information in the pack can then be used by the teacher to relate to current themes and areas of teaching. Information provided in packs should be factual and presented clearly, not dogmatic and consisting of a muddle of fact and opinion. Farms and farm centres can be promoted to educationalists by inviting them on free factfinding visits to the farm and showing them the range of resources available.

The educational market is unlikely to be lucrative, but it need not be a cost to the farmer. Small charges per head can be requested and expenditure on food and souvenirs will increase returns. The mark-up on biros and pencils with the farm logo will be much greater than that on the expensive craft products. The successful management of educational visits cannot, though, be measured just by the measuring rod of money.

Recreational enterprises on farmland are a far more varied bundle of possibilities than is the case with tourist enterprises. There is no doubt that they can be financially and socially rewarding. The potential for enterprises of this type will vary markedly from place to place. Accessibility will matter. The clientele may be holidaymakers or people travelling from their homes. The urban fringe can be as desirable a location as the fringe of a national park. These enterprises cannot normally be established without planning permission, except where infrequent events can be accommodated within the 28 day rule. For some activities, such as motor cycling events, the site would be exempt for only 14 days. However, the success of recreational enterprises cannot be guaranteed by location alone. An appropriate product mix that will attract and sustain visitor interest must be identified. Where this is achieved, the commercial potential of recreational enterprises may be considerable if sound management can be combined with a flair for marketing.

Increased demand for recreation in the countryside is a mixed blessing. The tendency for passive recreation to be replaced by more active recreation may increase the pressures on farmland without significantly increasing income earning possibilities. However, a farmer attuned to the trends is more likely to be in a position to develop profitable enterprises.

Some have rightly questioned the statistical evidence of change and have found it difficult to draw a clear picture of likely trends.

Plate 4.8   *Farm open days*
Few fail to be interested in newborn lambs which provide a suitable
focal point for a farm open day. It is not inconceivable to make
farm open days financially successful if catering and craft
enterprises can be grafted on.

Plate 4.9   *Working exhibits*
            A live museum is not impossible as Acton Scott proves with its
            working horses and butter making. Static displays are often much
            less interesting than working exhibits.

The difficulty is not, however, ascertaining whether increased
demand will occur, but in isolating in what places and in what
activities. Will increasing unemployment force certain people to
use the urban fringe rather than the more distant countryside?
Will the traditional countryside be increasingly peopled by affluent
urban refugees imitating the recreational pastimes of traditional
rural residents and adding a few new activity pursuits of their own?
These questions must be considered if the changing market for
recreation is to be explored and new opportunities are to be
exploited.

ADDING VALUE TO CONVENTIONAL FARM PRODUCTS

The possibility of adding value to conventional farm products before
they leave the farmer's hands has aroused considerable interest
in recent years.[23] Producers and representative organisations have

looked to the concept of added value as a means of salvation from the problems of the cost–price squeeze and the threat of lower prices for conventional products. However, whilst the principles of adding value might seem eminently sensible, there are pitfalls that must be recognised. The rationale and methods of adding value must be scrutinised.

The concept of adding value can be examined by looking at the humble potato (see Table 4.9). Sold in the wholesale markets, potatoes are cheap. Acquired at the farm gate in 25 kg bags, they are slightly more expensive. Sold in the market place in an unwashed loose form, they can be significantly more expensive. Washed and bagged, their price is still higher. Washed and labelled as baking potatoes in Marks and Spencer's, they are almost a luxury item. Finally, sealed into a cellophane packet with a Smiths or Golden Wonder label, they sell for thousands of pounds a tonne. The same principles are applicable to any farm product including livestock products or even the products of farm woodlands as well as fruit and vegetables: processing or alternative forms of marketing can add value (see Figure 4.2).

Table 4.9 Potential for adding value to potatoes

Form of sale	Price per tonne (£)
Wholesale	60
Retail, loose, unwashed	220
Retail, bagged, washed	280
Retail, pre-packed, washed, baking	940
Potato crisps	5250

Note: Prices prevailing in Spring 1989

The principle of value added must be refined before it can become useful. The concept of *gross* value added must be differentiated from that of *net* value added. The net value added must be rather less than the gross value added when the costs of adding value have been taken into account. These costs may include fixed costs such as additional full-time labour or new processing machinery, and variable costs such as casual labour. Existing resources on the farm should be costed at their *opportunity* cost (the earning power foregone by using them in the value adding process). Thus, under-employed family labour should be costed at zero cost if no additional wage is paid.

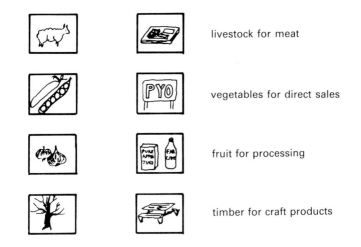

livestock for meat

vegetables for direct sales

fruit for processing

timber for craft products

Figure 4.2 The scope for adding value to farm products

Value is added by transforming a product but it is only realised when the product is sold. Pick-your-own does not add value to strawberries when they lie unpicked in the fields. The experiences of Curworthy Farm with its new cheese are a further indication of the problems of endeavouring to add value. With the product in its infancy, production problems created wastage, and market resistance to its high price reduced sales. Additional processing creates additional scope for errors. For a while, the Curworthy Cheese enterprise appears to have been reducing the value of the raw material rather than adding to it.

As adding value to a product normally takes the producer closer to the final consumer, the need for adopting a marketing approach, and the perils of neglecting it, are all the more apparent.

Value can be added to a product by the primary producer in two ways (or a combination of both). Firstly, alternative channels of marketing can be employed which yield a higher return than those used normally. This approach can be seen clearly in the development of direct marketing and the increment to producers' incomes from cutting out the wholesalers' margin. Secondly, it is possible to add value by processing a raw product. The conversion of milk to cheese or yogurt, of fruit to jam, and of meat to pies or pâtés are all examples of this. Value adding enterprises may pursue either or both of these courses of action.

ADDING VALUE BY MARKETING

There are many alternative routes between the points of production and consumption and attempts to add value by alternative marketing strategies have followed a number of main lines.

- Produce can be delivered to the customers' door by the producer. This is commonly found in milk and in vegetable production.
- Produce can be sold from the farm gate, at stalls or in farm shops. All vegetable crops, meat and dairy products can be sold in this way.
- Produce can be sold at PYO enterprises. These apply only to horticultural enterprises and merit separate attention.
- Produce can be marketed through conventional market centres – either from market stalls on an occasional or regular basis or permanently from a farm shop – but the farmer is involved in the whole marketing chain.

All of these methods eliminate or reduce the need for middlemen to be involved in the marketing process. Consequently, an understanding of marketing principles and techniques is a fundamental requirement for a farmer indulging in such methods of adding value.

It is dangerous to regard wholesalers and middle men as parasites and undesirables. They may possess knowledge and skills which will enable them to find a market for a product. They may be attuned to the needs of the supermarkets. Furthermore, they may be able to ensure regularity of supplies by aggregating the output from a number of producers. These middle men may be private firms or producers' co-operatives. The producer should review the options and act accordingly.

Many of the methods of adding value to farm products are poorly documented. Farm gate sales were often seen as the perks of the farmer's wife rather than an income earning opportunity that merited study. However, in the wake of quotas and uncertainty relating to most major food products in the EC, alternative marketing has become the subject of much greater attention.

Door-to-door Sales

An unknown number of farmers is involved with the production and retailing of vegetables by door-to-door sales. This method of adding value is common in northwest England where there are many small farmers producing horticultural products close to large

Plate 4.10   *Direct marketing (1)*
The sign is large and visible, the buildings attractive and the parking straightforward. All these are important ingredients in direct marketing. The sign could, however, be improved. The gaps and the mixture of upper- and lower-case letters do not make for easy reading for the passing motorist.

Plate 4.11   *Direct marketing (2)*
Attempts to combine premiums on organic products with direct marketing make sense. A good location on a main road with adjacent layby is not helped by bad signs haphazardly thrown together.

conurbations. The advantages to the farmer are clear. He can sell his produce at competitive retail prices, obtaining significant margins over wholesale prices. The location of the farm is much less critical. Farms that are inaccessible to main roads but still close to major urban areas are able to benefit from proximity by lower delivery costs. The additional revenue that can be obtained must be set against any additional costs which may include a delivery van, washing and bagging equipment and additional labour costs. There is also a danger that the production aspects, such as timeliness of field operations, may be neglected. These hidden costs must also be considered. Where family labour can be pressganged to help and costs can be kept low, there are undoubtedly opportunities. However, the consumer can acquire fruit and vegetables in many different ways and although many consumers do not have the opportunity to purchase fruit and vegetables on the doorstep, there is no evidence to suggest significant potential for growth.

An intermediate strategy which might be adopted by fruit and vegetable producers is to approach retailers direct and avoid the wholesale markets. If producers are able to find short cuts in the marketing chain and deliver good-quality produce to retail outlets like corner shops and small supermarkets, it is possible to offer the consumer competitively priced and fresh fruit and vegetables that compare very favourably with products of larger supermarkets. A similar strategy can be adopted with eggs or with milk and milk products. As in the case of door-to-door sales, proximity to the market is desirable but a prime location on the transport network is unnecessary.

In the livestock sector, the most common form of involvement with door-to-door marketing of produce is with milk rounds. There are, in 1988, approximately 2600 producer retailers and produce processors in the UK. Although this is only about 25 per cent of the number of producer retailers in 1960, the volume of milk sold in this way has declined by much less. This decline is symptomatic of the extent to which milk producers had lost touch with the market place prior to the imposition of quotas.

In order to become a producer retailer or a producer processor, who processes milk and delivers it to retail outlets, a licence must be obtained from the Milk Marketing Board. Producers have been deterred by the intense competition that has come from the large dairy companies which, even if they do not directly operate milk rounds, will supply independent roundsmen with competitively priced milk and a wider product mix than can normally be offered by a producer retailer. Furthermore, there has been uncertainty about the future of the doorstep delivery of milk in the face of increased competition from supermarket sales of milk. There is, though, something intrinsically logical about satisfying a local demand for liquid milk (and milk products) from local supplies. Distribution costs must be saved even if other economies of size are not always achieved. Where the balance would be in a perfect market can only be guessed, but producer retailing and producer processing may not be as anachronistic as recent history has made them appear.

There is little evidence of farmers getting heavily involved in direct marketing of meat by taking the product to the consumer. The one area where farmer involvement has grown is in supplying farm-produced meat to the consumer, often for home freezers. Livestock can be sold to slaughterers and carcases bought back or the slaughterhouse paid for killing the stock without its being sold. In either case, the farmer obtains a carcase which must be butchered professionally if it is to satisfy the consumer. Private arrangements can normally be made with those with appropriate skills. The extent of the practice is constrained by the need for butchering skills, the capital investment required and the multitude of regulations that must be adhered to. There are two main approaches adopted by producer retailers in order to differentiate their products from other types of sales outlet. Firstly, they can compete on price and a casual examination of local paper advertisements normally yields evidence of competitively priced meat for the freezer direct from the farm. Secondly, the scope for home deliveries may be considerable where there is a quality/speciality label on the product, such as oak smoked ham or organically reared lamb, the latter being offered for sale (and sold) in summer 1985 at twice the price of normal lamb.

It is normally assumed that a market location is essential for an enterprise involved in direct marketing of food. However, there are other ways of direct marketing, which include mail order, which are potentially important ways for the less well located farmer to reach the consumer direct. It is possible to use very rapid delivery services which ensure the maintenance of quality of the product. Thus, a food-smoking firm in the West Highlands of Scotland or an organic meat producer in southwest England can both market direct to the Home Counties even if they are not located on the consumer's doorstep.

Farm Gate Sales

Farm gate sales, either through roadside stalls, through farm shops or simply from the farmhouse door, have a long tradition but have been refined and extended in recent years. The small-scale sale of eggs from the farmhouse door represents one end of the spectrum, whilst the other is represented by the pseudo-rustic food supermarkets in the countryside that masquerade as farm shops. Where food is sold at the farm gate, it is obviously desirable for it to be located close to a large town or city and on (or visible from) a major traffic route from which customers can exit readily. It is not impossible to operate farm gate sales from a less advantaged location, for personal recommendations can be a major creator of custom. However, where locations are disadvantageous the marketing is even more crucial. The farmer must be able constantly to improve his service to customers as competitors endeavour to imitate his successes. As competition with this type of enterprise intensifies, the disadvantages of a sub-optimal location may increase.

Where farm shops are established, the farmer immediately encounters a range of regulations. With mobile stalls and existing buildings used to sell farm-produced fruit and vegetables, planning permission is not needed. If additional products are bought in or if meat processing and retailing is carried on at a farm, planning permission must be sought. There is also a need to adhere to weights and measures, food hygiene and health and safety regulations. The full range of regulations is outlined in *Farm Sales and Pick-Your-Own*.[24] The National Farmers' Union have recently published a very thorough guide to farm gate sales to the public which is essential reading for anyone entering this field.[25]

Farmers trying to attract the public to farm gate sales must recognise that normally they will have to persuade the public to travel further to obtain a product that could have been acquired much closer to home. Great importance must therefore be attached to the image conveyed to the public. Food may be sold on the basis of competitive pricing, higher quality, or freshness and wholesomeness. Farmers should recognise that for many 'shoppers' there will be a recreational element to the visit to the countryside and should be conscious of maintaining an attractive presentation of their products in an attractive setting. The challenge to the farmer is to differentiate his product successfully from what can be obtained in a conventional shop. Given the interests of the consumer in countryside as well as food, the farmer should be in a position to achieve this if the ground rules of marketing are adhered to and marketing techniques successfully applied.

Plate 4.12   *Farm shops (1)*
Whether or not the eggs come from a battery unit, the contrived
image of the farmyard hen is cleverly conjured up with the wicker
basket.

Plate 4.13   *Farm shops (2)*
Effective presentation in a farm shop. The 'shop' consists of a
multi-span greenhouse and has allowed what started as a PYO
enterprise to extend its season and still provide goods on wet days.

Plate 4.14   *Farm shops (3)*
              Direct sales of meat products offer a way of adding value by
              marketing and processing. The Heath-Robinson style sign may get
              around planning laws but it is important to promote the right image
              to the customer.

Pick-your-own Sales

Pick-your-own enterprises are a much more recent development
in the UK than farm gate sales. Their numbers grew very rapidly
from a handful at the outset of the 1970s to nearly a thousand by
1985. They are, predictably, concentrated in traditional horticultural areas, especially those close to the London and Midlands
conurbations. This rapid expansion has generated a certain amount
of research and even official benediction in the form of a socio-
economic advisory leaflet and ATB courses. In addition to examin-

ing the type of PYO establishment and its location, some of the research quoted by Bowler[26] has explored the reasons why consumers are attracted to PYO establishments. Firstly, consumers are looking for competitive prices, frequently for products that will be bought in bulk to fill their domestic freezers. Secondly, freshness and quality of the product are considered important; finally, a visit to a PYO farm is likely to be seen as a recreation as well as a shopping expedition by the shopper. Pick-your-own enterprises cannot be considered as a last ditch effort to shift poor-quality produce to the consumer. Whatever the initial reasons for entry into PYO, the contemporary manager must be closely aware of the consumers' wants and recognise the highly competitive nature of the business he/she is in.

Many growers who entered PYO relatively early have diversified and adapted their enterprises in various ways. Children's play areas, snack bars and toilets have been provided and more conventional farm shops have been added for those customers unable or unwilling to pick their own produce. Further value can be added by converting surplus products into jams and preserves which can then be sold in the farm shop.

The evidence on the profitability of PYO enterprises indicates that they can generate highly variable returns. Of the published evidence, only that of Sangster[27] and the *Farmers Weekly* Marshcroft Farm[28] is directly comparable. Sangster estimates that the incremental margin from PYO exceeds that for farm gate sales of a conventional nature and that PYO gross margins can be between 160 and 300 per cent of normal gross margins. The figures for Marshcroft Farm are actual rather than estimated and although for 2 years later, show lower gross margins in all products (see Table 4.10). The differences can be explained partly by crop failures but, nevertheless, suggest that desk estimates of profitability may be rather higher than can be obtained in reality. Regional differences in markets and in crop potential must be taken into account and any generalisations about profitability regarded with caution.

As is the case with farm shops, PYO enterprises must comply with regulations relating to weights and measures, etc. Where ancillary enterprises are added, planning permission is likely to be necessary. Managers should pay particular attention to issues of accessibility and to signboards, both of which are likely to be closely scrutinised by officialdom.

Table 4.10 Gross margins from PYO enterprises, estimated and actual

	Sangster Northeast Scotland 1980 (£)	Farmers Weekly Hertfordshire 1982/83 (£)
Strawberries	3062	1479
Raspberries	3182	2105
Potatoes	1663	1141
Carrots	2450	663

A healthy outlook for PYO cannot be guaranteed. It seems most improbable that the consumer interest in PYO and related enterprises will wane, though it may be necessary for managers to alter their product mix over time. The ideal products to sell are those that have high picking costs, such as mangetout peas, but that does not mean the whole unit can be put down to one crop. In an increasingly competitive climate, it is essential to identify a range of products that capture the imagination of the public and to market these products efficiently. The full repertoire of marketing skills is likely to be put to the test from pricing to promotion, to getting into a new product a season ahead of the competition. Future opportunities will be highly dependent on location and the biological potential of the ground as well as the abilities of the manager. It is impossible to generalise about the prospects for two principal reasons: firstly, the degree of competition; secondly, the individual's location. Each case must be treated on its merits.

Normal Retail Sales
The final opportunity to the farmer for adding value by marketing in alternative ways is to establish retail premises in a normal market centre. Many market towns contain shops and stalls which are part of a wider farm business. Close links with the farm are frequently used in promoting the produce sold from the shop. These retail outlets are subject to the same range of regulations as any other food retailing outlets and those relating to meat are especially stringent. Farmers running such shops will need to consider their ability to supply the retail outlet with the requisite products continuously. Consequently, those that exist tend to sell products not subject to seasonal production, such as poultry, pork and dairy products. It is important to recognise the high levels of fixed costs associated with high street or even side street shops. Where market

stalls are rented, fixed costs will be much lower and the problems of continuity of supply are much less. However, the occasional use of retail markets is unlikely to generate the customer loyalty and regular trade that retailers seek to gain.

It is impossible to generalise about the financial returns to these various forms of direct marketing. A survey of farmers in northwest England published in 1987 indicated that 20 per cent of farmers had on-farm marketing operations.[29] Many forms of marketing were used, as is indicated in the marketing of one product: eggs (see Table 4.11). Prices up to 79.7p/dozen were obtained compared to the normal wholesale price of 48.2p/dozen. The author notes that these figures should be treated with caution, but they are sufficient to point out to farmers the possibilities for adding to the price received. At a time when so much additional expenditure on food is given to others in the food chain it is both consoling and instructive to see such opportunities manifested.

Table 4.11 Relationship between marketing method and egg price

	Pence per dozen
Normal wholesale outlets	48.2
Average: all on-farm marketing	53.0
Range of prices	43.6–79.7
Farm-retailed eggs	60.9
Eggs sold direct to trade	57.7
Eggs retailed from delivery round	79.7

Source: Russell.[29]

The future of the marketing approach to adding value lies in the hands of the consumer as well as the producer. The evidence of the last decade shows two trends which are scarcely complementary. The rapid growth in the market share of hypermarkets and supermarkets with all types of food suggests that there is little hope for the small producer retailer. However, lurking in the shadows is another type of consumer, willing to use the supermarket for convenience shopping, but with an interest in locally produced fare which is tastier and fresher and seen as more 'natural' than the offerings of the supermarket. The importance of this market segment is seen in the attempts of supermarkets to imitate the qualities the customer seeks. The evidence of this is seen in the colour supplement advertisements and in the shops themselves. Real ale campaigners took on the giant brewery companies and created major changes in policies as well as creating new opportunities for small producers. The same opportunities may be there for producer retailers of food products where, for so long, so many have predicted impending doom.

ADDING VALUE BY PROCESSING

Processing of farm products provides an additional way of adding value that can frequently be used in a complementary manner to adding value by alternative marketing strategies. On-farm processing may produce a final or intermediate product which is normally a food, but may consist of a product such as straw briquettes. Processors range from traditional clotted cream producers in the back lanes of Devon to large highly entrepreneurial food processors like Baxters of Speyside who still retain their links with the Morayshire soil. Somewhere between these extremes are the medium-sized farm cider producers who have retained their traditions for sufficiently long for their products to be drawn into the same net as the real ale revival.

There is a pitifully small amount of information about the extent and profitability of on-farm processing. In spite of the June returns which reveal the whereabouts of every hen and sheep, there are no official data relating to on-farm processing of food or other products, or to the profitability of such enterprises. Generalisations about profitability are fraught with difficulty. Data about the decline in farmhouse cheesemaking must be set alongside press reports of farmers complaining of quota restrictions that make them unable to satisfy a growing demand for their product.

Such generalised evidence as there is suggests that 'the market for wholesome country foods and drink is expanding'.[30] This judgement seems to consist of marginally more that an act of faith, for press reports indicate an interest in such foods as far apart as Cumbria and Devon. The potency of rustic images in food and drink advertising reinforces this assertion but the notion of 'wholesomeness' may be, at best, vague and, at worst, misleading. Can foods containing large amounts of animal fats, often disguised in pâtés or pies, really be conducive to bodily or spiritual health? This would certainly not be the view of many contemporary commentators on the links between diet and health. In spite of these contentious areas, or perhaps because of them, discerning consumers may prefer to consume their reduced intake of animal fats in the form of quality farmhouse cheese rather than the mass-

Plate 4.15   *On-farm processing (1)*
This display of goods on sale in the farm shop indicates how on-farm processors have adapted. Sheppy's have also encouraged visitors to view cider making and visit a museum of cider and farming.

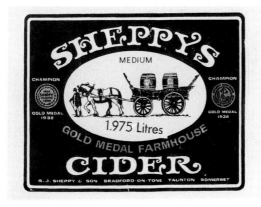

Plate 4.16   *On-farm processing (2)*
This label for farmhouse cider bears a logo which is found on most of the products in the farm shop. There is also evidence of Weights and Measures problems. The '1.975 litres' has been stuck on to the label subsequently, replacing '2 litres'.

produced alternative. It is easier to identify stoneground wholemeal flour or farm-processed apple juice as wholesome country food. Farmers should not adopt a dismissive attitude to the health issues, for health education is likely to advance consumer awareness much more in the imminent future.

A small but significant group of food writers have been promoting quality food for some time. Jane Grigson of *The Observer* and Jeremy Round of *The Independent* have an intuitive feel for quality, and their interest in food extends back to the production process on the farm. Cheeses like Bonchester and Lanark Blue from Scotland and Heal Farm products from Devon have been praised, and such praise must benefit those marketing the product. It would be misleading to assume that these writers represent more than a minority of those interested in food. Nevertheless, the views they promote to a small but growing following should give hope and stimulation to the farm processor of food. The innovation of food products is not solely the domain of the biochemist's laboratory; it can equally take place in the farm kitchen.

Processing Livestock Products

The processing of meat is surrounded by stringent regulations and frequently requires considerable skill and capital. Slaughtering must be carried out at a licensed slaughterhouse but carcases can be returned for on-farm butchering or processing. The simplest level of processing involves no more than the jointing of meat into appropriate cuts for the direct sales trade. Secondary processing, which converts meat and offals into meat products such as sausages, haggis or hams, often use traditional farmhouse skills and do not require expensive and extensive equipment. Low value meat can be converted to high value meat products, but there is no guaranteed market for the products and market research must precede production. Public houses, restaurants, cafes and delicatessens might be approached, all of whom might benefit from the country food image. However, quality control and the ability to maintain supplies will be important considerations whether the outlets are shops or farm gate sales. There is nothing necessarily agricultural about these enterprises. In the past, such meat processing was a routine part of rural life. If, however, the farmer or his spouse can produce the inputs cheaply, if buildings are available and if someone is motivated by the idea and able to sell the product, 'traditional' quality foods may offer opportunities as alternative enterprises.

Milk is perhaps the most amenable livestock product for on-farm processing. At its simplest level, the production of green top (i.e., untreated) milk requires no more than a bottling plant, though the threats to ban sales of such may eventually succeed in eliminating this option. The next level of processing is the pasteurising plant which gives the farmer processor a wider range of marketing options. There is also scope for on-farm processing of milk into higher value products like cheese, yogurt or ice-cream. The capital demands and technical skills required for soft cheese and yogurt are rather less than those required for hard cheese production. It may be possible to initiate production of dairy products to traditional local recipes. Consumers are likely to be interested in low fat products and this should be borne in mind at the planning stage.

There are difficulties with processing milk because of the many regulations that affect the UK dairy industry. Producer retailers and producer processors must be licensed by the Milk Marketing Board. All producers are also affected by quotas and after the introduction of quotas, there was a time when a farmer with wholesale and retail quotas was unable to transfer between them. This absurd situation has thankfully been resolved. Milk producers can apply to the Milk Marketing Board for schemes which permit a change of quota from wholesale to direct sales. There is one loophole in the quota scheme which allows anyone with a direct sales quota to produce yogurt and dairy ice-cream on quotaless milk. In theory, this should offer opportunities for on-farm processing to those who have, or can obtain, a direct sales quota. How long this anomaly will remain can only be guessed.

The dairy sector in the UK has been perhaps the most innovative in terms of diversification in recent years. Quotas shocked producers into reviewing their options, and the limitations imposed by quotas closed the conventional route of coming to terms with the cost–price squeeze. The new cheeses and other dairy products that have been developed represent one of the adjustment pathways adopted. That these innovations should be a reaction against quotas rather than demand-led is a cause of concern, for the protectionist measures that still give them a reasonable return from milk cannot do the same for milk products.

A further obstacle to be surmounted in the case of adding value to milk is ascertaining whether processing and selling milk products from the farm gate requires planning permission. At least one farmer has had his business activities trimmed by planning authorities who deemed that direct sales of ice-cream to the public were causing an environmental nuisance in a Yorkshire village. This example indicates how important it is to be clear as to what planners will and will not deem to be acceptable changes of use with added value enterprises.

As well as meat and milk products, value can be added to live-stock products such as eggs by converting them into cakes or lemon cheese. Both of these uses may provide a suitable use for cracked eggs. Other animal products may also be sold after value has been added. Skins or wool, particularly of less common breeds of live-stock, may offer value adding opportunities. It may be necessary to allow the processing to be carried out off the farm in order to guarantee a high standard of work. Tourists or individuals or groups interested in craftwork are potential customers.

Processing Crops and Vegetables
The classic example of on-farm processing of cereals is the pro-duction of flour by traditional mills. Examples have been cited in the farming press of profitable wholemeal flour production from farm milled wheat. Old mills may provide a visitor attraction and farm gate sales can be promoted by this means. The buoyancy of the health food market should be recognised. Other examples of farm processing of cereals might include the production of crushed oats for horses or the use of mill and mix plants for neighbouring farmers. The possibility of processing straw merits serious attention in view of the antagonism that straw burning arouses and the demand by powerful pressure groups for a complete ban. If straw briquettes can provide warmth to the urban refugee's country cot-tage there may be a triple benefit: his warmth and the absence of a pall of smoke to set against the value added in processing to the farmer.

The scope for adding value to vegetable products is greatest where washing and packing facilities are available and these are unlikely to be on the farm of the smaller producer. Fruit products can be converted into jams and jellies. Although the demand for jam has fallen, there may be opportunities for producing farmhouse-style jams which offer a pleasant contrast to the products found on the supermarket shelf. If such activities are to be carried out commercially, then food regulations must be adhered to and atten-tion given to market outlets. Otherwise the village harvest festival may be the beneficiary of large amounts of unsold products.

Fruit can also be converted into alcoholic or non-alcoholic bever-ages. Cider sales rose substantially in the early 1980s but this has now levelled off and cider producers are fighting to retain their market. Nonetheless, the observations of *The Sunday Times* taste panel which 'voted for gutsy real ciders with rustic qualities and a high alcoholic level'[31] should not pass unnoticed. Apple juice is being produced in various parts of the country at a farm level, although it must be recognised that the equipment is not cheap and the technical skills needed may be considerable. Other fruits may have potential for processing, sometimes, as in the case of fruit yogurts, in combination with dairy products.

The processing of farm products demands two sets of skills. Firstly, there is a much greater requirement that attention be paid to marketing than is the case with supported farm products. Markets must be thoroughly researched and the full range of skills relating to the marketing approach must either be bought in or acquired. Secondly, there is a need for skills of a more technical nature. These vary considerably from the high-level skills required for butchering or cheddar cheese production to the basic skills of operating a pasteurising plant. Profitability comes from producing something that the public wants and can acquire through appropriate outlets at a price that satisfies them and yields the producer a reasonable return. The farmer's advantages in adding value are likely to be cheap inputs and premises and the growing interest in traditional products. These advantages can be derived only if new skills are acquired and developed.

UNCONVENTIONAL AGRICULTURAL ENTERPRISES

The term 'unconventional' is used to describe those enterprises which are not major agricultural activities but which involve animal or crop production. As the markets for most conventional products stagnate or decline, so a broader array of prospects is being considered. The line between 'conventional' and 'unconventional' is inevitably arbitrary and is very much a product of local practice. Thus, in some parts of the US, hog production and potato production are regarded as alternatives. Viticulture is unconventional in the UK, conventional in France. Some crops are frequently referred to as alternative crops in the UK where 'alternative' is a synonym for break crops. Such crops as peas, beans and rape are not seen as alternatives in the context of this book. They are merely alternatives to cereals, but are mainstream agricultural crops.

There are two principal categories of alternatives: those that are at or near market and those that still require substantial research to ascertain whether they are likely to be adopted widely. None of the established alternatives will provide, in national terms, a major substitute for conventional agriculture enterprises. However, it is possible that alternatives currently being researched, particularly

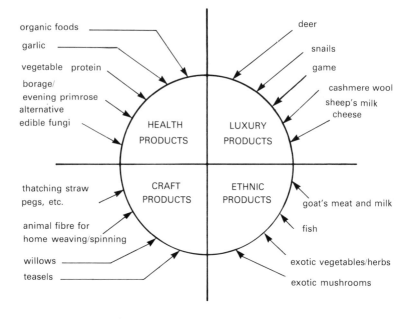

Figure 4.3   Unconventional agricultural enterprises

those linked into the rapidly developing field of biotechnology may, at some future time, create a major demand for land. New industries may be developed to refine vegetable products into fuel, fibre and chemicals. As these are not enterprises that the individual farmer could be expected to develop at present, they are excluded from the following analysis, although it is recognised that at some future point it may be necessary to reappraise their prospects at a farm level.

A thorough analysis of all unconventional enterprises is impossible but it is possible to structure the alternatives in such a way as to recognise the markets to which they belong. A representative selection of alternatives is offered in Figure 4.3.

A casual perusal of Figure 4.3 should be sufficient to indicate certain groupings of these unconventional enterprises. Four main groups can be identified and some products fall into at least two. These groups are chosen because they reflect certain significant market trends. The first consists of products which are seen as beneficial for health. It includes vegetable proteins, organic foods and such crops as borage and evening primrose. A second group consists of luxury items including farmed game, sheep's cheese, snails and cashmere. The third consists of products of interest to ethnic minorities: vegetables such as okra, herbs such as coriander, and goat's milk and meat all appeal to ethnic minorities whose buying power is by no means insignificant. Crops for export fit into the same category. The final group of products relates to the craft revival. Thatching straw can be grown; animal fibres can be produced, teasels are still used for fine-quality cloth. Examples will be taken from each of these areas for illustrative purposes.

HEALTH PRODUCTS

Health products can be broken down into two subgroups. Some of the products are destined to become foods which are, to a greater or lesser degree, health foods. Others are destined to be processed and used in pharmaceutical or homeopathic applications such as borage or evening primrose. Much of the thinking behind organic food production relates to health but also includes somewhat different philosophical issues.

The relationship between food and health has become, and is likely to remain, a topical issue and one that will alter dietary habits. The principal concerns relate to the assertions that we as a nation consume too much fat, especially animal fat, too much sugar and too little fibre. Some of these are the concern of the processor but there are also significant implications for the producer. Consequently, low-fat meats or non-animal sources of protein are likely to interest the consumer of the future. Both goat and deer have very low levels of fat in the carcase compared to lamb (see Table 4.12). An alternative approach is to replace animal by vegetable proteins and fats. Attempts to cultivate the navy bean (the baked bean) have yet to indicate sufficient potential for commercial production, but plant breeders and seedsmen are looking to alternative crops to broaden the range of products on offer. Attempts to develop sunflowers suited to UK climatic conditions are being made, spurred on by the expansion of demand for sunflower oil and realisation that large quantities are imported.

The massive attention given to *Salmonella* and *listeria* in 1988/89 in the UK cannot be dismissed as an exclusively British concern. Concerns about food and health are equally prevalent in continental Europe and in the US. Health aspects are likely to be potent factors affecting the markets for different foods throughout the developed world.

Table 4.12 Composition of goat, deer and sheep carcases

	Goat (%)	Deer (%)	Sheep (%)
Lean (%)	59.8	72*	49.8
Fat (%)	10.1	15	25.1
Bone (%)	22.7	12	19.2

*including some bone

Sources: Goat and sheep data from Wilkinson and Stark[32]
Deer data from Blaxter

Ironically, the greatest interest in vegetable proteins has come from the search for soya substitutes for feed compounders. Crops like lupins are likely to be used increasingly in livestock feeds. Its growth within the EC is a function of protectionist agricultural policies which now offer subsidies on lupins. The possibilities for lupins are likely to be restricted climatically but profits may be obtained in the short term as astute growers play the crop rotation game according to the distorted rules of the CAP.

Borage and evening primrose have both been advocated as alternatives in recent years. Both contain gamma-linolenic acid which is in demand for the treatment of medical conditions ranging from multiple sclerosis to premenstrual tension. Both crops are still in their infancy and, of the two, evening primrose is the most likely to develop, albeit in a small way. Early trials work identified problems with crop establishment, weed control and harvesting. More recent trials have looked at planting methods (fine seedbed versus transplanting), plant spacing, fertiliser applications and the performance of different cultivars.[33] As an alternative break crop, evening primrose is likely to be grown by a relatively small number of farmers. Whatever its medical reputation as a panacea, it is unlikely to be a panacea for cereals producers seeking alternative crops.

A number of crops contain agents which reduce levels of cholesterol in the blood. Gamma-linolenic acid is one such agent. Garlic is reputed to have similar beneficial properties. Trials have taken place in southern England which indicate yield possibilities of 6.25–7.5 t/ha and gross outputs well in excess of £5000/ha at 1985 prices.[34] Certain edible fungi that are widely consumed in the Far East and are likely to be grown in Britain have similar effects. Lentinus edodes, the Japanese wood mushroom, is being grown in West Germany, Poland and North America. In addition to the ethnic demand for this product, it is likely to interest the health food market segment, for it benefits health in a variety of ways.

Plate 4.17 *New crops: the Japanese wood mushroom* Lentinus edodes
This mushroom is widely consumed in the Far East, revered by
gastronomes and being actively developed in the US. MAFF is
not interested in funding research on it.

Organic agriculture has its origins in alternative environmental
philosophies rather than in the market place. But there is more
than a little evidence to suggest that the Green values, long extolled
by organic farmers, are being taken up by more and more people
and a few politicians. Thus, the philosophical logic of organic farming might combine with the market logic of expanding demand,
which might combine with the policy logic of finding land-using
activities to reduce surpluses. Although certain segments of agribus-
iness find it difficult not to belittle organic farming and continue to
endeavour to repudiate the claims of its supporters, organic farmers

have something that agribusinessmen have not got: a supporters' club that is growing in size.

The principal interest of consumers in organic food is in the perceived healthfulness of the product. The concern about harmful residues in food that is produced under intensive production systems has led many people towards organic foods. Secondly, many consumers believe that organic food has more flavour. Thirdly, those that ponder the future of the planet are of the opinion that, in general, organic production systems are rather more environmentally benign than conventional farming. Finally, those concerned with the creation of employment in rural areas argue that organic farming requires additional labour and is, therefore, better at sustaining rural communities.

In general, a distinctly less dismissive attitude to organic farming prevails in the late 1980s compared with 5 years ago. The first supermarket to stock organic food was Safeway, and now between 3–5 per cent of the greengrocery sales come from organic produce. Sainsbury's have carried out test marketing and now stock organic food. A Safeway buyer has estimated that if regular supplies could be guaranteed, the market share of organics could rise to as much as 10 per cent. The sorting out of organics standards, nationally and internationally, is fundamental to the development of the market and has recently been addressed in the UK by various committees of the UK Register of Organic Food Standards.

The increasing interest in organic production has extended to universities and colleges. The first major piece of research on the economics of organic farming and a review of the policy implications was carried out in the early 1980s.[35] The Royal Agricultural College now runs an organic unit amongst its farms and organic agriculture is now almost respectable in academia. Policy analysts are looking at organic production as one means of extensification, but to grant-aid intensive farmers into organic production could undermine the price premiums that established organic producers feel is necessary to support their production systems.

Value can be added to organic products by processing. The Doves Farm label of organic flour is produced at a farm mill in Berkshire and retails in shops and supermarkets with a 15 per cent price premium. Jordans have developed a range of products using what they term 'conservation grade' cereals. Conservation grade is a halfway house on the road to full organic certification and is viewed with suspicion by fully organic farmers, for it is likely to eat into their market. Organic milk is advertised as an ingredient in an ice-

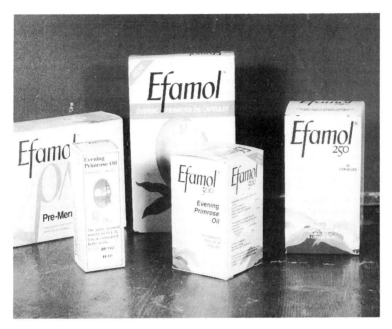

Plate 4.18   *New crops: evening primrose*
Pharmaceutical and health products are the outlet for evening
primrose oil. It is not an easy crop to grow and until production
systems have been perfected, the potential health gains to
consumers may exceed the potential financial gains to farmers.

cream produced by a Devon farmer. All these examples indicate
the scope for adding value to organic products by processing.

It would be naive to see organic farming as a means of solving
the excessive support costs of the farming industry and, in addition,
all farm-related problems of environmental pollution. Such evidence as there is points to the fact that incomes on organic farms
do not compare favourably with incomes on conventional farms.
But, bearing in mind many of the early organic farmers were adopting their systems for non-financial reasons, it is somewhat irrelevant
to measure their success with financial indicators. Substantial
premia can be obtained for organic food and much of the UK market
is currently supplied from overseas. Co-ordinated marketing and
cropping programmes, perhaps influenced by co-operative marketing agencies, could greatly enhance the development of this sector.
The early difficulties faced by some of the organic co-operatives
seem to be largely behind them and some have experienced a

dramatic growth in turnover in recent years. Environmentally, there are grounds for believing that organic systems are likely to be relatively beneficial compared to intensive conventional farming systems.

The principal challenge to the conventional farmer contemplating conversion to organic production is the transition period, when low yields will conspire with low prices to create cashflow problems. For these reasons farmers are advised to move piecemeal towards organic husbandry. Although the purists may find this opportunistic, financially oriented approach to organic farming somewhat distasteful, they should be consoled by the marginal benefits to the environment that are likely to result. However, as the cost–price squeeze on conventional production intensifies, and if consumer attitudes continue to shift towards favouring organic products, this sector can be expected to develop significantly in the next decade. Market research evidence substantiates and justifies the interest of the supermarkets.[36]

Advocates of organic farming stress that it should not be seen as a diversification enterprise but as a way of managing whole farming systems.[37] However, as farmers move piecemeal into organic farming to ease themselves through the difficulties of the transition period, they are likely to be practising organic farming as an alternative enterprise. The organic movement is keen to see farmers adopting organic farming over the whole farm. In time, it may be that organic standards will require whole farms to be run organically in order to obtain the organic symbol for marketing purposes.

LUXURY PRODUCTS

The market for luxury foods depends on the level of wealth of the affluent minority which buys these products. The aggregate wealth of a nation may be less important than the distribution of wealth. The market for venison is unlikely to have been dramatically affected by the recession in Scotland; the demand for caviare in the Soviet Union is not a reflection of the overall wealth of the population of the country. An expansion in the buying power of the affluent minority, either by increasing its numbers or increasing its wealth, is likely to increase the demand for luxury foods. This market has been recognised and the farming of game, particularly deer and the production of expensive cheese including that from sheep's and goat's milk, are an indication of how producers have responded. Other examples include snail production which has been described as 'a slow way to make a quick profit',[38] while the *Financial Times* noted the growing demand for snails in Italy 'in certain restaurants for the better off'.[39] Wine produced in England may not be regarded as a luxury beverage by the connoisseur but it is a further reflection of producers' attempts to create foods and drinks for a more affluent populace.

Snails must represent one of the most hyped-up, but unsubstantiated, livestock alternatives. It seems that nothing can go wrong. The French market is waiting to swallow up anything that can be offered. Experts selling their snail stories to a gullible press also have expertise at selling their packages to prospective snail farmers. Can there be a catch? If snails are so efficient at reproducing, why have not the French devised their own means to expand production? It is most un-French to allow the English to develop expertise in an expanding French market. Caution must be counselled. There is no background of research into snails in the UK. Production systems remain in their infancy and are as yet unrefined to meet the needs of particular situations. The gourmet restaurant market is limited but not without some potential and the export market likely to yield much lower returns than direct marketing.

Deer farming is another example of a luxury product that has received a great deal of press attention. Pioneering research was conducted on deer husbandry by the Rowett Research Institute and at Glensaugh in northeast Scotland. In the late 1970s and early 1980s the deer farming industry looked likely to boom. It did so in New Zealand (see Chapter 1) but has never taken off in the same way in the UK. Although the boom has failed to materialise, the interest in deer has not disappeared. By the mid 1980s there were about 100 deer farms and 150 deer parks in Britain,[40] although only about 50 units were considered to be 'commercial' deer enterprises in 1987.[41]

The failure of the industry to expand as had been predicted can be attributed to many factors. The recession may have influenced the demand marginally, but more important was the failure to back up husbandry research with financial evaluation. The capital costs of setting up a deer farm are considerable. Fencing costs are between £3.00 and £3.75/m, and the stock is significantly more expensive than sheep or cattle. Although high gross margins look favourable, the higher fixed costs make the benefits questionable. Furthermore, there is no agreement as to what levels of gross margin can be expected. In 1987/88, gross margins per hind on lowland intensive systems selling venison from stags and breeding hinds ranged from an estimated £70, to nearly £150, per head.

These differences require close scrutiny, for if interest rates are high they are easily sufficient to turn a high return on capital into a significant loss.

The marketing of farmed venison has been relatively weakly developed. One commentator has noted that a critical mass has yet to be reached, unlike the situation in New Zealand.[42] Once this has been attained, new markets are opened up, especially in the up-market supermarkets. Those who have established the deer farmers' marketing co-operative are conscious of the need to provide regular supplies of a consistent product to larger retail outlets. To date, small-scale producers have found it relatively easy to market direct and obtain good returns which means that, for many, there is little incentive to use the co-operative.

It would be premature to dismiss deer farming as a prospect. Early researchers provided much sound background material, which is a model of how such early development work should be conducted.[43] However, if the high establishment costs, especially in relation to fencing, are recognised, it makes sense to look at alternatives to the hill and upland deer production systems. Two possibilities present themselves. Firstly, it may be worthwhile to consider deer on lowland farms where higher stocking densities can be achieved and fencing costs per hind are less (and there is evidence that this is occurring). If researchers had paid more attention to the red deer's low-ground origins rather than its present upland distribution, the lowland possibilities might have been more fully explored at an earlier date. However, it may be necessary to take account of red deer behaviour with regard to breeding and feeding in the layout of red deer enterprises in the lowlands. Secondly, it may be possible to manage red deer populations in a wild state on their adopted habitats in the hills and uplands. Winter feeding of wild deer may improve survival and growth rates. The deer can remain wild and the value of a stag can effectively be doubled by having someone paying to shoot it. There are likely to be no fencing costs to contend with in this production system.[44] It is unfortunate that the Highlands and Islands Development Board persisted in trying to establish financially viable extensive deer farming in the late 1970s when a strategy based on improved range-land management and stalking might have been more rational. Furthermore, if the demand for breeding stock exists, it may be rational to contemplate domestication of Scottish hill stock for there is widespread evidence that wild red deer numbers are too high. In the event of a hard winter a very high level of mortality would

Plate 4.19   *New livestock products: venison*
Farmed venison production has increased. The product has many
positive factors: low fat, luxury image, etc. but it remains to be
seen whether it is a rich man's hobby or a significant alternative
enterprise for the average farmer.

occur. It may be better to move some of the deer down the slope
to deer farms rather than let them starve on the hillsides.

In spite of evidence that margins are higher for deer than for
sheep or cattle in the uplands, the evidence on profitability is
ambiguous, largely because of the high establishment costs. The
products of deer farming are unsupported by headage payments or
end price support which will be passed back up the hillside. If hill
and upland policies are reformed and less money is available,[45]
especially for larger hill farms, the rangeland management option

may become more profitable. The prospects for profitable farming are likely to be greatest where deer farms can be established in the lowlands and venison can be sold direct to final consumers or catering establishments. As a luxury, low fat product, venison is likely to be sought after. The challenge remains to producers to create systems which produce reasonable returns.

Another livestock product reputed to be experiencing increasing demand, particularly in the export markets, is sheep's milk and related products. There are now about 350 milking flocks in the UK compared with a mere handful 5 years previously. Many established herds are expanding and new enterprises are being developed all the time. Milk production from sheep is not new. Thomas Tusser, the author of a famous sixteenth-century treatise on agriculture, wrote:

> Yet manie by milking (such heede they doo take)
> not hurting their bodies much profit doo make.

in his advice on tending sheep.[46]

Sheep's milk production is at the mercy of market forces. There are neither fixed prices for milk nor quotas to contend with. The producer must find his own market for the product although producers' organisations offer advice to members. Sheep's milk can be used in various specialist cheeses or sold as liquid milk. It can be frozen and readily stored. The market for sheep's milk is considerable in the Mediterranean area and for ethnic minorities in Britain. Processors have turned sheep's milk into yogurt and cheeses which are sold through a variety of outlets both to ethnic minorities and to high-class food shops like Harrods. A highly praised Roquefort substitute is being produced in southern Scotland. In spite of optimism amongst sheep's milk producers, the producers' association wisely advises members to ensure that they have markets for their products prior to establishing enterprises.[47]

It is quite impossible to generalise about the profitability of sheep's milk production. The recent expansion of the industry appears to reflect some considerable potential. However, the greatest beneficiaries of this boom are likely to be the owners of existing milking flocks who can sell their limited numbers of milking sheep at high prices to prospective producers. Unless reliable export markets can be developed, the scope for expansion in the industry is likely to be limited. There will be increasing competition in the delicatessen market from a growing range of sheep's, goat's and cow's milk cheeses, but it must be recognised that it is the speciality cheese market that has shown the greatest growth. The health food market is limited in total size but may grow if allergy sufferers prefer sheep's, to goat's, milk.

The opportunities for producers of luxury foods are unlikely to diminish. There is, though, no guaranteed market for luxury products and where the luxury market is limited the dangers of oversupply must be recognised. In many cases there is no formal market in the sense of wholesalers accustomed to dealing with a product. The marketing initiative must be taken by the producer or in some cases by producers' organisations. Private individuals and public organisations have recognised this gap and, in view of the growing demand for luxury products, are endeavouring to respond. However, the producer who can market his product direct is at a distinct advantage. Sales to hotels or from the farm gate may command a premium over wholesale prices in so far as they exist and accessibility to markets is a relevant consideration. Geographical remoteness may impose a significant cost with certain products.

A further example of a luxury item that has witnessed a remarkable expansion coupled with narrowing margins and increased competition is fish farming, particularly the freshwater farming of trout and, more recently, the farming of salmon in the Highlands of Scotland. These two fish dominate the fish farming industry, but the scope for coarse fish farms producing fish for human consumption, for restocking or for ornamental purposes, should not be overlooked.

Trout consumption in the UK has grown dramatically, increasing by 300 per cent between 1977 and 1982 although, by 1982, trout still accounted for only 2 per cent of fish consumed.[48] The fish farming industry has taken off after a development period lasting more than 10 years, but it remains to be seen at what point the expansion will be limited by a combination of oversupply, site limitations and narrowing margins.

Trout farming can take place with two potential markets in mind: that for table trout and that for restocking. The table trout market consists of several segments which must be analysed by the farmer, for the price of the product is likely to vary markedly from one segment to another. The farm gate market commands prices around £1.35/lb. at 1989 prices. Direct sales to restaurants can yield £1.20/lb., whilst wholesale prices can be under £0.90/lb. In the early 1980s, about 25 per cent of production was sold through direct sales and the remainder to wholesalers and processors. The advantages of direct sales are clear, but they are unlikely to be obtained by the larger production units.

Plate 4.20   *New livestock products: trout*
Trout farms can be part of a larger farming business but are often
self-contained. This example is part of an estate. It also adds value
in various ways, e.g. by charging visitors and selling pâté.

Earlier work by Lewis[49] showed significant economies of size in
rainbow trout production with smaller units experiencing significantly higher fixed costs per unit of output. However, smaller units
may be able to recompense themselves by a greater amount of
direct marketing. There has been a marked trend towards direct
marketing in the industry over the last decade.

Trout farming requires a suitable location in two respects. Firstly,
it is clearly advantageous to be located so as to enhance the prospects for direct marketing. Secondly, a precondition for trout farming is water of a suitable quality, with adequate dissolved oxygen
and freedom from pollution. The alkalinity of water should lie
between 20 and 200 mg/litre of calcium carbonate and pH should
be in the range 6.5–8.0.[50]

Fish farming is neither exempt from all regulations, as are many
agricultural enterprises, nor is it as constrained by many regulations
as certain alternatives on farmland. Fish farming itself does not
require planning permission as long as it is exclusively for food.
However, water authority regulations control the creation of ponds

or dams, the discharge of effluent and, in some circumstances, the abstraction of water. Consultation with water authorities is essential. Grant aid may be paid on fish farm developments under EC-initiated grant schemes and by regional development agencies such as the Highlands and Islands Development Board.

Trout can be farmed under a variety of systems which are fully described in standard texts. The needs and interests of fish farmers are also serviced by two trade journals and many vocationally specific courses in universities and agricultural colleges. Trout farming enterprises cannot be effortlessly tacked on to existing farms, unless adequate attention is paid to the physical requirements of the site, the technical skills required of those running the enterprise and to the ability to market the product to maximum benefit. Latecomers to this area of production will need to depend upon an extremely attractive location and/or a high degree of entrepreneurial flair in what has become a very competitive business. New ideas for adding value to the product should be considered, including processing into smoked fish or fish products, inviting the public to see and perhaps feed the fish for a fee, or running a small intensive fishery where the skill of catching a fish is limited to the ability to hold a rod.

A final example of a 'luxury' crop is viticulture. Where vines are grown as an alternative crop, wine making normally occurs too. There are now 546 ha of vines grown in the UK but of these, 163 ha are not yet in production.[51] Viticulture and wine production outside the main production areas are often regarded as little more than a rich man's hobby. The heavy capital investment required and the significant wait before the grapes are harvested and the wine sold go a long way to explaining this label.

Many wine experts have highly praised wines produced outside conventional production areas. New Zealand wines have been the most recent recipients of such praise in the UK. Some English wines, too, have been applauded for their qualities. In addition to producing for this quality market, there is a niche in the market for 'locally produced' products. These can appeal to the parochialism of local residents or the adventurous spirit of tourists.

Wine producers have become, like their colleagues in the whisky business, adept at promoting their ventures. In the US, wineries have developed in some areas only relatively recently and have often linked visitor attractions and farmgate sales to their development. In Santa Barbara Co., California, there are now 4400 ha of vines. Prohibition knocked out commercial wine production in the area, which was not restarted until the 1960s. Many of the wineries

**1987
OHIO RIVER VALLEY
VIDAL BLANC**

TABLE WINE - PRODUCED AND BOTTLED BY
LOUIS JINDRA WINERY
B.W. OH 298   JACKSON, OHIO   ALC. 12% BY VOLUME
CONTAINS SULFITES

Plate 4.21   *Luxury products: viticulture and wine making*
Wine production has expanded out of its conventional areas of
production. Often, in marginal regions, various strategies are
sought to add value to the product, including farm gate sales,
demonstrations, tastings, etc.

now offer guided tours and in some cases a significant proportion
of total sales is sold to visitors. Entry to wineries is free and the
visitor rarely departs empty-handed. This approach contrasts starkly
with some UK producers who charge for their guided tours.

In other non-traditional wine producing areas of the US, different
approaches to adding value to wine production can be found. In
southern Ohio, a wine producer has converted old farm buildings
into a restaurant as well as opening the vineyards and the winery
to the public. But these wineries are not the product of local
farmers' efforts at diversification. Brooks Firestone used wealth
acquired from the motor industry to develop the Firestone Winery
in Santa Barbara and the proprietor of the Louis Jindra Winery in
Jackson, Ohio, came from a professional background to develop his
much smaller enterprise.

Whatever the origins of the producers, lessons can be learned. Value is added to grapes in wine production. It can be added again by direct sales to the public, either over the counter or with a meal. These means of adding value exist for English wine producers too but have been more highly developed abroad. They actually provide sufficient returns to justify further enterprise development. There may be a case for a more co-ordinated approach to product promotion such as is seen in the whisky industry. Individualism may be fine for individuals, whereas a co-ordinated promotional strategy, including visitor attractions, may enhance the market for the product as a whole.

Luxury products need not be foods. Much of the current interest in goats centres on fibre production. Three types of valued fibre are produced from goats, with different breeds yielding different qualities as shown in Table 4.13

Table 4.13 Classification of various fibres produced from goats

Cashmere	- underhair	Below 19 μm diameter
Mohair	- Angora fleece	23 μm diameter and above
Cashgora	- from angora cross breeds	19–22 μm

Cashmere is the most valuable fibre and significant quantities are imported into the UK from politically unstable parts of the world. Work is currently underway at the Macaulay Land Use Research Institute to upgrade the cashmere content of feral goats. Not only are feral goats seen as potentially valuable for their fibre, but they also produce meat. It is possible to graze them with sheep and they will selectively graze-out vegetation not consumed by sheep. Indeed, this was the rationale behind their experimental introduction to the hill grazings at Glensaugh. Realisation of the potential of cashmere production with a market worth £70 million/year in the UK has led to a reorientation of research towards enhanced cashmere production.

Mohair is less valuable than cashmere and comes from an animal less well adapted to the UK climate. Angora production is currently expanding in Australia, New Zealand, Canada and the US and the scope for competition in the market place is considerable. The early interest in Angora often came from dairy goat producers who have been using Angoras to diversify their goat enterprises. The market for Angoras has recently experienced a market downturn and late entrants who paid high prices may be financially vulnerable as a result of the bursting of the bubble.

ETHNIC FOODS

Britain contains ethnic minorities of significant numbers which are often geographically concentrated. Long-standing cultural variations have been catered for by regional specialisms in food production. More recent cultural changes have yet to affect food production patterns significantly although there may be scope for producing certain products in the UK. The most obvious example of this is the production of goatmeat, milk and milk products which began its recent expansion in the hands of smallholders and has grown and adapted itself to a significant ethnic market. Other less well-researched examples include coriander production (as a green herb, not seed) and okra production under glass or polythene. The range of potential products also includes fish such as carp and alternative types of edible mushrooms. In addition to looking at ethnic influences on demand within the UK, overseas demand may be sufficient to justify British production. In all cases, success depends upon an ability to combine technical and marketing development work effectively.

Goat farming is moving rapidly away from its smallholders' origins and is now recognised as a farm enterprise with considerable potential. The transformation of the goat in New Zealand provides a model of how things might develop in Britain. Traditionally, goat enterprises in the UK have been dairy based, but there are three markets for goat products in fibre, meat and milk products. These markets are complex, including ethnic minorities for meat and some dairy products, and the luxury and health markets for dairy products.

The current interest in goat dairying owes a great deal to the introduction of dairy quotas. As goat's milk operates outside EC support arrangements, the market place, rather than Brussels bureaucrats, will determine the price for the product and will guide producers' behaviour accordingly. It is asserted that the market for goat's milk products is unsatisfied but it is important to consider the cross elasticities between goat's and sheep's milk products. They are undoubtedly competing in an overlapping market. Evidence from the mid 1980s suggests that goats can compete reasonably successfully with unquota-ed grazing livestock enterprises.[32]

Recent events in the development of the goat industry are informative. When an alternative enters a rapid expansion phase, optimism can be unbridled and absurdly high prices can be obtained. It is, of course, in the interests of those with breeding stock to create the hype. Those that fall prey to it are likely to have to struggle. Often these new entrants have gone into the industry with no real knowledge of the market possibilities. There is a strong case for market research for a broad array of goat products, parallelling that carried out for mohair.[52] Much of the needed research might be deemed near market by a parsimonious MAFF. But whose interests will be funding the market research? If it is the processors, they are unlikely to be playing down the market. The more producers, the greater the supply of the product and the lower the price paid to producers. Whether the price of the garment, fashioned from the luxury fibre, falls in value is a different matter, especially if there are near monopolies in the manufacturing market place.

The case of goats indicates that the markets for ethnic products are often interwoven with luxury markets. In the case of ethnic foods, this may well be the case, because of increased international travel. But the buying power of ethnic minorities within the UK should be recognised.

The range of ethnic minorities is considerable, but the two largest groups in the UK are Asians from the Indian subcontinent and West Indians. There are also smaller groups of Chinese and various immigrant groups from European and Near East countries. Frequently these groups are geographically concentrated which would make distribution of products relatively easy. The demand for different foods by different cultures and subcultures should lead farmers to think about their ability to satisfy overseas demand. In spite of trading problems with politically unstable countries such as were experienced with payments for Scottish-produced feta from Iran, there is an affluent Arabic market in the Middle East which can be considered. The obstacles to domestic production in Saudi Arabia make British environmental difficulties pale into insignificance.

CRAFT PRODUCTS

The potential for producing the raw materials for craft products on the farm depends very much on how strict a definition of craft products is adopted. Here, the term will be used to include crops like thatching straw, teasels, willows for basketwork and animal fibres for hand spinning, weaving and knitting. Furthermore, crops from farm woodlands may also be converted into craft goods from bird boxes to garden seats.

Plate 4.22　*New livestock products: goats*
There is a growing interest in goats for meat, milk and fibre. Finding
the right genetic material and the right production systems for
different products creates a new challenge, especially for meat and
fibre.

Plate 4.23　*New livestock products: goats for fibre*
Angora goats have been promoted for their fibre. Prices have
boomed and slumped in various parts of the world in recent times.

Plate 4.24 *New livestock products: wild boar*
Wild boar are farmed by a landowner/hotelier who can offer his
guests a different flavour of roast pork. It is more of a marketing
gesture than a major profit-making enterprise but if it increases
public awareness of the hotel it may have achieved its objectives.

The demand for craft products can be broken down into distinct
areas. Firstly, there is a demand for raw materials from practitioners
of crafts who are then selling a product to the final consumer.
Secondly, there is a demand from the recreationist pursuing a craft
or a leisure pursuit. As early retirement increases and employees
work a shorter working week and have longer holidays, it is likely
that the demand for craft activities will increase. Over time, the
distinction between crafts as a business and as a hobby is likely to
decline as those using craft activities as a leisure pursuit begin to
find markets for the products. Thirdly, there is a demand for craft
products in their finished form.

The craft industry consists of a highly varied set of firms including
many small businesses. Some craft firms are highly market oriented,

turning out almost mass-produced crafts, whilst other craftsmen are the antithesis of the market orientation, concerned only about the quality of their product, regardless of its saleability. The varied nature of this market must be recognised by anyone dealing with it. Whilst the more commercial, market-oriented firms will need continuity of supply of products, the demand from hobbyist craftsmen is likely to be sporadic unless products can be channelled through the appropriate outlets.

The growing interest in crafts can also be exploited by linking this recreational activity with the provision of tourist accommodation in activity holidays.

The packaging of such holidays and appropriate marketing strategies must be complemented by an ability to satisfy the needs of the public. Alternatively, buildings may be developed to offer craft workshops to interested craftsmen. These opportunities will be examined in the next section.

Many of the unconventional agricultural products are less well tested than conventional products. The essential ingredients in successful husbandry are less well known. There is (and should be) an understandable reticence amongst farmers to dive headlong into large-scale production of unconventional products. However, this necessary caution must be balanced against the scope for the farmer to adapt and develop production systems and obtain the premiums that can be obtained by those that adopt the new products at an early date.

ANCILLARY RESOURCES

Many farms include resources which, although they may have value in the economy in general, have little or no value to agricultural activity. Farm buildings may be rendered redundant by modern technology. Farm servants' or family rooms in large farmhouses constitute a potential tourist resource. Woodlands and wetlands may be useless in terms of agricultural potential but have a valuable alternative use. These alternatives have tended to be neglected while support for conventional products is high, but are likely to interest farmers more as this support is reduced.

BUILDINGS

The combined effect of specialisation and mechanisation have left a legacy of buildings which are considered important elements of the rural landscape but are partly or completely redundant. The agricultural options on these buildings may be extremely limited but they may have significant value for conversion to other uses. It is not always old buildings that become redundant. Relatively modern farm buildings become so through farm rationalisation or sales of farms in lots to neighbouring farmers. There are two broad classes of conversion: to residential uses, and to commercial uses. (It may be convenient to separate out tourist uses from other commercial activities.)

The demand for these different categories of use is likely to vary dramatically from region to region and significantly within regions. Thus, in 1989, a barn in South Devon can be valued at £70,000 or more with outline planning permission, whereas a steading in rural Aberdeenshire might be worth, at best, £15,000. There are also large disparities in the value differences for tourist use but the variations in other commercial values are much less. These differences are obviously influenced by regional variations in house prices but also reflect the price that people will pay for amenity either for residential, or recreational, use.

The expansion of demand for barns and the increasing numbers of conversions is evidenced in data collected from planning departments in a number of English districts[53,54] (see Figure 4.4). Most of these conversions of farm buildings are for residential purposes. The incidence of conversions is such that, in some Devon parishes, there have been more conversions than registered agricultural holdings.

The demand for rural residences appears to have strengthened over the 1980s. The extension of the motorway system and major road improvement schemes have pushed residential demands into more rural districts. The planning system has tended to operate a policy of restraint in rural areas with higher levels of restraint operating in the highest amenity areas. In national parks and green belts, new developments are likely to be resisted but conversions might be allowed. This has given barns a premium value. The planning system, by limiting the supply of new homes, operates as a fairy godmother to the owner of buildings with conversion potential.

The conversion process to residential use can be an expensive undertaking. Where the process of conversion is protracted, cash-flow problems can arise. There can be delays in letting or selling the accommodation even when completed. Frequently, the farmer would be advised to sell the barns having obtained outline planning

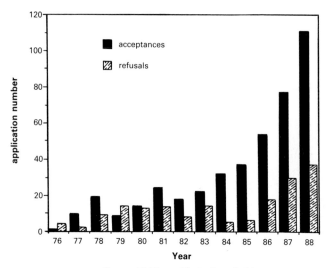

Source: Mid Devon District Council 1989

Figure 4.4   Applications for barn conversions in mid Devon 1976–88

permission, rather than retain a share in the conversion and carry an increased burden of risk.

The encouragement of small firms in rural areas has long been a component of rural policy in the UK. The Rural Development Commission (formerly CoSIRA, the Council for Small Industries in Rural Areas) and equivalent bodies in other parts of the UK have had such functions since the early years of the present century. The RDC's efforts are concentrated in Rural Development Areas which are designated on the basis of a range of socio-economic indicators. Within these extensive areas, grants are given for conversions (except to residential uses) on a discretionary basis, at similar rates to those operating under the Farm Diversification Grants but without the relatively low ceiling. In addition, in certain circumstances the RDC will offer loan finance. In areas like the Highlands and Islands of Scotland and mid Wales, regional development agencies fulfil these industrial development functions. As well as the work of the RDC, central government has actively promoted small firms throughout the economy. Advisory services are offered

Ready for conversion

THIS exceptional range of stone barns for development has just come on the market. They are at Budleigh Barton, close to Compton, a small village between Newton Abbot and Totnes, and only five miles from the centre of Torquay.

They are an outstanding example of traditional farm buildings with beautiful stonework, arched openings, slated roofs, and incorporating a superb roundhouse.

Extending to almost 13,000 sq ft, they are in a 'U' shape and surround a courtyard area which has immense potential. Planning permission has been granted for change of use to five residential units, the conversion of which would, of necessity, be to a high standard with finished dwellings expected to realise in excess of £100,000.

The buildings are in a convenient, yet delightful rural setting, and land between one and two acres is available to the purchaser's requirements.

Offers are invited in the region of £120,000, and the agents are Stags, of Totnes.

Plate 4.25(a)   *Barn conversions (1)*
The possibility of capitalising on the value of farm buildings depends on the whims of planners. Where permission is given, the income gain can be considerable.

Plate
4.25(b)

Barn conversions (1)
The same barns as in Plate 4.25(a), now converted. One of the
five units was recently on the market at over £200,000.

Plate 4.26   *Barn conversions (2)*
These new buildings are now used for farm work in place of the
barns in Plates 4.25(a) and (b). Elegantly landscaped with the
occasional exotic conifer, they await the interest of a twenty-first-
century yuppie.

Plate 4.27   *Barn conversions (3)*
These redundant oasthouses have been converted into permanent
accommodation. The urge to escape from the monotony of a
Bovis, Wimpey or Barratt home puts a premium value on
accommodation like this.

from central government such as the small firms service of the
Department of Trade and Industry.

The demand for industrial premises will vary from one part of
the country to another. For many commercial uses of buildings, a
rural location is an option, not a necessity. Consequently, an attractive setting is advantageous. It may also be advantageous where, in
the case of craft workshops, customers may be coming to the place
of production to purchase the products. Accessibility and parking
availability can also be important factors in such circumstances.
Often it is the isolated buildings that farmers are most willing to
release. Not only may these be inaccessible, but they may not have

mains services and these can be very expensive to lay on. Rentals for buildings for industrial use vary from £1.75 to £4 per square foot per annum. It may be necessary to allow low levels of rent to attract clients in and, although it may benefit the small business-man, it can create cashflow problems for the farmer.

Many industrial developments to date have been carried out on larger farms and estates. This may be a reflection of the supply of buildings, often released from agricultural use as a result of farm rationalisation. Equally, it may reflect other factors, including the availability of capital to see the development through, or the aware-ness of land agents of the market opportunities. Even on smaller farms, opportunities may exist, but if converted buildings are close to the farmer's residence he should, wherever possible, retain their ownership.

Barn conversions to tourist uses are inevitably concentrated geo-graphically but it should be recognised that there are possibilities outside traditional holiday areas like Devon, Cornwall and the Lake District. Typical shire counties may also have potential, as may areas close to London. However, a number of observers fear satu-ration levels may have been reached in some districts. Early entrants in an area who establish themselves in less than optimal locations may find that they are progressively squeezed by later entrants into the field.

Those who have taken advantage of the Tourist Board grants in the past have been obliged to use an approved letting agency as a condition of getting the grant. However, in 1989 the Tourist Board grants were abruptly curtailed. Anyone converting buildings must trade-off the higher costs of agency marketing against the larger letting period that they can normally achieve. Once established, return visits and personal recommendations should be a major source of custom. An absence of this source of trade should raise questions about the value and quality of the accommodation offered.

Anyone seeking to convert an agricultural building must recog-nise the constraints and opportunities afforded by the planning system. Although normally regarded by farmers with a mixture of contempt, suspicion and hatred, it may, in the case of barn conver-sions, be the farmer's salvation. The planned scarcity of develop-ment opportunities creates a premium on such buildings. However, planners can also constrain, and a minority of applications is rejected. There is growing concern that conversions may threaten the buildings which are among the more interesting remnants of vernacular architecture and planners are entirely reasonable in asking developers to respect the buildings that they wish to convert.

Plate 4.28   *Barn conversions (4)*
The conversion of this barn in north Buckinghamshire has retained
the character of the barn yet created an attractive home.

Plate 4.29   *Barn conversions (5)*
Barns can provide industrial premises as well as homes. These
barns in Herefordshire have been converted to builders'
workshops.

Plate 4.30   *Barn conversions (6)*
This attractive barn within Milton Keynes has been converted to
craft workshops.

Plate 4.31   *Barn conversions (7)*
Farm buildings do not need to be large to have potential for conversion. This conversion of cow sheds in Oxfordshire is covered by a Section 52 agreement which ties it to agricultural ownership and prevents any possibility of the farmer deriving its open market value.

Plate 4.32   *Barn conversions (8)*
This barn has been converted to a tearoom and craft centre, grant-aided by both the Tourist Board, for the tearoom, and the RDC for the workshops.

Plate 4.33   *Farm woodlands (1)*
Many farm woodlands are neglected. This woodland has been
replanted with oak using Tuley tube tree shelters.

One major concern that has recently been aired by a number of
commentators is the change in the General Development Order
which makes planning permission obligatory for new livestock
buildings or facilities for storage of slurry or sewage sludge if they
are within 400 m of residential buildings.[55] Thus, farm development
in the future may be constrained by objections from those who
have moved into a converted barn. Those contemplating barn conversions should recognise these risks, and also that future changes
in planning policy – such as the release of rural land for housing –
might reduce their value. The lack of understanding of the planning
system needs to be conquered if rational developments are to be
possible in the working countryside without unnecessary bureaucratic impediments.

WOODLANDS AND TIMBER

There are thousands of small woodlands on farms, many of which have received little management for decades. This neglect is not without some financial justification. Two world wars robbed the lowland woods of much of the timber of any quality and over a longer period, cheap coal and hydrocarbons and the longer-term replacement of woodland craft products by mass-produced goods have created little financial incentive to manage woodland. Only game management and occasionally the demand for fuel or small roundwood generated a demand for management. Some woodlands suffered benign neglect. Others were grubbed out to add acres to agricultural enterprises. In addition, the lack of a forestry tradition on farms in the UK is cited as a reason for the neglect of woodland resources.

A combination of factors has begun to reverse this trend. There is a growing recognition of the amenity and conservation value of trees, especially of ancient woodland. This is reflected in action from the public, the voluntary, and the private sectors. The urban 'refugees' living in the countryside are interested in woodland to enhance their amenity and fuel their woodburning stoves. After decades of neglect and decline, farm woodlands are being actively explored as income-generating possibilities for farmers, and funding and other forms of advice are being promoted by agencies like the Forestry Commission, the county councils, and voluntary bodies like Silvanus.

The demise of broadleaved woodland was also a function of the dominant strategy of the Forestry Commission. This created the coniferous forests of the uplands and inevitably research tended to focus on conifers. The resultant pattern of afforestation created much antagonism in hill and upland communities. However, where integrated land management schemes have been devised, as in Glenlivet in northeast Scotland, the partial afforestation can be seen to benefit the upland farmer by creating shelter and access. In other instances, the release of capital for afforestation in the hills has allowed investment in land improvement such that stocking rates can be enhanced rather than reduced. It is more difficult to make a case for integrated management where the prevailing farm type is small owner-occupied upland farms, but in the areas of more extensive hill sheep farming, partial afforestation on farms may constitute a plausible strategy both to farmers and land use planners.

In various countries in the southern hemisphere, agroforestry practices are being developed and refined. Under agroforestry systems, the objective is to use the same area of land for two purposes: a tree crop and, normally, grass production for ruminant livestock. New Zealand researchers have led the way in developing these systems and there is a developing research interest in the UK.[56]

In recent years, broadleaved woodland has attracted increased interest. The Forestry Commission produced a major manual on broadleaved woodland in 1984[57] and followed this in 1988 with a publication on farm woodland planning.[58] The proliferation of books on farm woodland planning is ensuring a continued demand for at least one wood product: paper. These texts provide useful guidelines for farm foresters.[59] One simple innovation has been the development and improvement of the Tuley tube, a tree shelter that dramatically improves the rate of early growth in certain trees, notably oak.

Grant aid for tree planting has changed recently with the introduction of the Farm Woodland Scheme. Hitherto, woodland development had been grant-aided by establishment grants paid in three instalments under the precursors to the 1988 Woodland Grant Scheme. The Farm Woodland Scheme gives annual payments of up to £190/ha/year to farmers for periods of up to 40 years. The grant is directed primarily at taking land out of agriculture and the high rates of grant are paid only for land coming out of arable or rotational grass. The scheme is running for an experimental 3-year period up to 1991, with an aim of producing an additional 36,000 ha of woodland. It is important to note that planting grants identical to those of the Woodland Grant Scheme are offered in addition to the annual payments. These grants consist of payments of up to £1375/ha.

The Farm Woodland Scheme is an attempt to expand the area of broadleaved woodland on lowland farms. It does not encourage the rehabilitation of existing woodland, although there is widespread evidence of neglect and undermanagement. There are provisions in the Woodland Grant Scheme to enable the revitalisation of derelict woodland either through replanting or managed natural regeneration. Under the 1988 set-aside provisions, grants can also be obtained for woodland planting, but under the set-aside grants, the payments will only be made for the duration of the set-aside scheme. For this reason, farmers would be advised to use the Farm Woodland Scheme wherever possible. It should be noted that all schemes require prior approval by the Forestry Commission. A final significant source of grants is the Agricultural Improvement

Scheme which offers grants for shelter belts and hedgerow tree planting.

Forestry policy is still in a state of flux. The Government's 1988 Budget removed forestry from the income tax system. Payments derived from the Farm Woodland Scheme (the annual payments) will be subject to tax, but payments from the woodland products will not. The immediate effect of the 1988 Budget was to reduce dramatically the planting rates of coniferous trees. High income earners who had previously found a tax shelter in forestry investments could no longer do so. Planting rates are likely to increase from their present low levels as private forestry companies attract new clients, but there can be no doubt that the fiscal changes have ushered in major changes into the structure of forestry ownership. Furthermore, there is some evidence[60] that returns to forestry may be rather better on medium grade land than on the poorer land which has historically been afforested. Although current land prices make afforestation on better land non-feasible, a further reduction in agricultural land values could make lowland silviculture a worthwhile option, especially if future grant schemes offer additional encouragement to farmers for planting trees.

Woodland management on farms can take many forms. It may consist of rehabilitating broadleaved woodland, planting-up wet areas with poplars or willows, or steep banks with broadleaves or conifers. Alternatively larger-scale coniferous ventures might be contemplated on hill farms or on light thin soils. Shelter belts can be planted and small woodlands can be established to enhance game value. Christmas trees will yield a return in a relatively short period of time. In the future there may be a case for planting-up parts of the lowlands with trees for energy cropping, possibly on a short rotation basis. The range of possibilities has normally been sufficient to create despair rather than interest but the rapidly shifting climate of opinion is likely to lead to greater support for woodlands in general and farm woodlands in particular.

The problems of making woodland owners aware of the silvicultural potential of woodland is not unique to the UK. British farmers are often compared with those in Scandinavia where there is more forest management and greater understanding of silviculture in the farming community. However, in the US, public and voluntary organisations are concerned about poor woodland management and a weak understanding of the marketing of timber.[61] There is a danger that selective felling of quality timber will lead to its replacement by genetically inferior woodland from seed sources left standing. Some commercial extractors plunder the woodland with little

Plate 4.34   *Farm woodlands (2)*
The notion that there is a choice between stock and woodland is
refuted in agroforestry systems. This Welsh Border woodland of
poplar has been grazed by stock with no evident damage to trees.

heed for the future generations. These poor forest management
practices and the need to create additional awareness of tree management in Conservation Reserve Program plantings signify a
demand for woodland management education and training in the
US as well as the UK.

Ironically, the demand for timber and the contemporary estimates of returns from woodland management cannot be used to
justify this new enthusiasm for trees. The estimation of future
demand for trees is fraught with difficulty. Few would hazard a
guess at the demand for oak when 1989's acorns are harvested
in the year 2119. The establishment of long rotation broadleaved
woodland has always been an act of faith, for uncertainty and high
discount rates are always likely to discourage such developments
in the absence of public support. The returns from woodland on
lowland farms rarely compare favourably with contemporary farming returns except on poorer land, but in the uplands it is possible

to find many instances where, at present prices, forestry returns exceed those of farming. Diminished agricultural support and increased woodland grants are likely to shift the advantage towards woodland but the likely extent of the shift can only be guessed.

The limited long-term economic justification for new woodland creation does not imply that there is no financial case for instituting some kind of management on some of the existing farm woodlands of the UK. There are markets for firewood, hardwood pulp and small roundwood, although prices offered are likely to differ substantially from one part of the country to another. If farm woodlands contain any high-quality timber this can command high prices for furniture making. An appraisal of markets and of silvicultural management needs is a prerequisite for re-establishing woodland management. Sufficient examples exist to prove that programmes for the restoration of derelict woodland can be profitable, although

Plate 4.35 *Farm woodlands (3)*
The starting point can be bare ground. Tree shelters are again used to promote effective establishment. The resultant woodland strip in southern Scotland will provide shelter and game value. Not all farmers can afford to take arable land out of production in this way in the absence of greater incentives for forestry.

much depends on the level of dereliction and the volume of timber that can be felled.

Christmas trees have frequently been advocated as a form of short rotation forestry suitable for farms. These enterprises produce returns relatively quickly and are readily comparable to agricultural investments. Making a profit from Christmas trees is highly dependent on marketing. The most profitable Christmas tree enterprises are those that can sell the trees to the public from the forest gate. In some years there has been oversupply of trees with very low wholesale prices. The emergence of specialist growers will inevitably pose a threat to new entrants who are unable to sell their product direct to the public. Thus, while Christmas trees should not be rejected as a possibility, neither should they be seen as a means of silvicultural salvation.

The major change in public interest in broadleaved woodland has created a significant policy shift. But for many farmers in the UK there is still a large step to be taken from the current practices of robbing derelict farm woodlands for fuel, to managing woodlands as commercial undertakings. Changes in public attitudes will need to be backed up by appropriate education and training to help those managing the land to seek out the silvicultural possibilities. In most cases, forestry enterprises will not be justifiable in terms of the financial returns in a free market. But the various lures and incentives should be sufficient to waken farmers to the opportunities that are becoming more and more worthy of investigation as agricultural returns decline. Thus the current enthusiasm of MAFF and the National Farmers' Union must be tempered by financial realism but farmers would be advised to improve their silvicultural understanding. As some cynics have observed, timber is one of the few primary products produced within the EC that is not in surplus.

WETLANDS

Wetlands exist on many farms. There is a wide variety of wetlands ranging from natural or artificial lakes and ponds to marshy areas. These areas have little agricultural value. They are, however, of interest in a number of other respects. Conservationists have shown great concern about disappearing wetlands, more often by writing about them than visiting them. This concern has resulted in nature reserve acquisitions, Sites of Special Scientific Interest (SSSIs) and management agreements, the most notable of which covers a large area of the Havergate Marshes in East Anglia.

Plate 4.36   *Wetlands*
A wet area of land has been converted to a low-cost trout lake.
Even in northeast Scotland with all of its river game fishing, there
is a buoyant demand for fishing this lake.

In the event of a management agreement not being forthcoming, farmers can look to other ways to exploit wetlands, by planting water-tolerant trees or by converting what is occasionally wet to something that is permanently wet by creating farm ponds. These ponds or lakes can be stocked with fish and provide an income from acres that formerly contributed little or nothing to conventional agricultural production.

Whether the farmer is looking at agriculturally redundant buildings or land, the same underlying principles should apply. He should recognise that he can offer a wider range of products to the public than conventional agriculture produces. Indeed, there may be a case for shifting resources out of agriculture into alternatives. Complete redundancy of existing buildings or acres is not a prerequisite for their consideration for alternative use.

PUBLIC GOODS

Public goods are those goods which have no financial value to the farmer in the absence of some contribution from the public purse. Economists define public goods in a formal manner as goods/serv-

Plate 4.37   *Public goods*
Much of the area in the middle ground is covered by a Management
Agreement under the Wildlife and Countryside Act 1981. Whether
the land was really likely to be reclaimed is a moot point. 'Farming'
grants may be more profitable than farming sheep.

ices from which no one can be excluded and where one person's
consumption does not adversely affect another's. In reality there
are goods and services which are more or less public. As rural
policy develops to take on board more of these public goods, so it
is incumbent on the farmer to be aware of the policy instruments.

Since the 1981 Wildlife and Countryside Act, compensation has
been payable to landowners for foregoing their rights to develop
land on Sites of Special Scientific Interest or on open moorland in
national parks. These compensation payments implicitly recognise
the value of the land for conservation of landscape or wildlife. The
levels of payment are based on the agricultural income foregone
and the critics have pointed out that the farm incomes foregone,
low as they may be, are still substantially supported by the EC. An
alternative procedure would be to compensate on the basis of prices
that could be obtained in a free world market for the product in question.

The introduction of Environmentally Sensitive Areas (ESAs) in
1986/87 brought a different approach. In all, 738,000 ha were designated in a wide variety of areas ranging from the hills of Breadal-
bane and mid Wales to the heaths of Breckland and the wetlands
of the Somerset Levels and the Broads. In each of the eighteen
areas designated as ESAs, schemes were drawn up for paying
agreed rates of grant for farming in an environmentally sensitive
manner. These schemes are administered by MAFF. Land or buildings of outstanding importance to the nation may be exempted from

Plate 4.38 *Diversification of diversification*
This farm in mid Buckinghamshire offers a farm shop, a caravan site and ballooning facilities. It is still quite clearly a farm even if it does not have the quaintness and rustic appeal of the farm of the imagination.

inheritance tax and capital gains tax. In this way, the landowner is compensated for maintaining the resource, but provisions must be made for enabling public access.

The final form of compensation payment is the access agreement – initially introduced in 1949, and subsequently broadened in its scope. Although this scheme is guaranteed not to make the farmer's fortune, it is a mechanism whereby compensation is paid for allowing the public on to private land.

The proliferation of compensation payments of various types in recent years is an indication of the extent and importance of these public goods to the population at large. There has been much discussion over whether or not farmers and landowners merit compensation for providing and maintaining resources with value for landscape or conservation. There is a powerful argument that compensation is justified, although there is room for disagreement as to appropriate levels of payment. Farmers must learn to live in this changing policy climate and recognise that they may be able to obtain benefits from farming in an environmentally beneficial way.

REFERENCES

FARM TOURISM AND RECREATION

1. GASSON, R. (1986) *Farm Families with other Gainful Activities.* Wye College.
2. PEAT MARWICK MITCHELL (1986) *Study of Outside Gainful Activities of Farmers and their Spouses in the EEC.* Commission of the European Community.
3. DALTON, G. E. and WILSON, C. (1989) *Farm Diversification in Scotland.* North of Scotland College of Agriculture.
4. LLOYDS BANK (1988) *Farming Issues,* 3. Lloyds Bank.
5. CARRUTHERS, S. P. (ed.) (1986) *Alternative Enterprises in UK Agriculture.* Centre for Agricultural Strategy.
6. BRITISH TOURISM AUTHORITY (1988) *Tourism Fact Sheets: West Country,* BTA.
7. MEDLIK, S. (1982) *Trends in Tourism, World Experience and England's Prospects.* English Tourist Board.
8. WEST COUNTRY TOURIST BOARD (1980) *A Strategy for Tourism in the West Country.* West Country Tourist Board/English Tourist Board.
9. ENGLISH TOURIST BOARD, RURAL DEVELOPMENT COMMISSION (1987) *A Study of Rural Tourism.* ETB.
10. WRATHALL, J. (1980) 'Farm-based holidays', *Town and Country Planning,* **49**, 6.
11. FRATER, J. (1982) *Farm Tourism in England and Overseas.* CURS Research Memorandum 93, University of Birmingham.
12. RIMES, R. (1984), op. cit.
13. WINTER, M. (1983) *Where Town and Country Meet.* Paper presented to the Rural Economy and Society Discussion Group, London.
14. TRADE AND INDUSTRY COMMITTEE (1985) *Tourism in the UK.* HMSO.
15. DART/RURAL PLANNING SERVICES (1974) *Farm Recreation and Tourism in England and Wales.* CCP 83, Countryside Commission Publication 83.
16. DAVIES, E. T. (1983) *The Role of Farm Tourism in the Less Favoured Areas of England and Wales, 1981.* University of Exeter.
17. WINTER, M. (1983) op. cit.
18. VEBLEN, T. (1934) *The Theory of the Leisure Class.* Random House, London.
19. WARREN, M. F. (1981) *Riding Establishments in the South West.* Unpublished Report, Seale Hayne College.

20. COUNTRYSIDE COMMISSION (1977) *Farm Open Days*. Advisory Series Publication No. 3. CC.
21. BLUNDEN, J. AND CURRY, N. (eds) (1988) *A Future for our Countryside*. Blackwell.
22. COUNTRYSIDE COMMISSION (1987) *A Compendium of Recreation Statistics*. C.C.

ADDING VALUE

23. ROSENTHALL, P. (1981) *Agricultural Production and Marketing in Cornwall and Devon. Recent Trends and Possible Developments*. University of Exeter Agricultural Economics Unit.
24. GROWER GUIDE NO. 4 (1979) *Farm Sales and Pick Your Own*. Grower Books.
25. NATIONAL FARMERS' UNION (1987) *Farm Gate Sales to the Public*. Shaw & Sons.
26. BOWLER, I. R. (1980) *Direct Marketing in British Agriculture*. University of Leicester.
27. SANGSTER, H. (1982) 'Farm gate sales with particular reference to northeast Scotland', *Farm Management Rev.*, **16**.
28. *FARMERS WEEKLY* (1983) 4 November.
29. RUSSELL, N. (1987) *Added Value Activities on North West Farms: results of a pilot survey*. Manchester University.
30. DEVON COUNTY COUNCIL (1985) *Food Distribution: its impact on marketing in the 1980s*.
31. *SUNDAY TIMES* (1982) 7 February.

UNCONVENTIONAL PRODUCTS

32. WILKINSON, M. AND STARK, B. A. (1982) *Goat Production*. Ministry of Agriculture, Fisheries and Food.
33. DODD, M. AND SCARISBRICK, D. (1987) 'Evening Primrose', *Crops*, 13 June.
34. TOMSETT, A. (1985) 'Garlic: a possibility for Devon', *Devon Horticulture in Focus*, May.
35. VINE, A. AND BATEMAN, D. (1981) *Organic Farming Systems in England and Wales*. University College of Wales.
36. *THE OBSERVER* (1986), 2 March.
37. LAMPKIN, N. (forthcoming) *Organic Farming*. Farming Press.
38. *GUARDIAN* (1979) 4 April.
39. *FINANCIAL TIMES* (1982) 22 December.
40. FARM ANIMAL WELFARE COUNCIL (1985) Press Release, January.
41. FLETCHER, J. (ed.) (1987) *Deer Farming: a realistic alternative?* British Deer Farmers' Association and Barclays Bank.
42. FLETCHER, J. (ed.) (1987) op. cit.

43. BLAXTER, K. D. (1974) *Farming the Red Deer*. HMSO.
44. MUTCH, W. E. S. (1978) *Red Deer in South Ross*. University of Edinburgh.
45. MCEWEN, M. AND SINCLAIR, G. (1984) *New Life for the Hills*. Council for National Parks.
46. TUSSER, T. (1931) *Five Hundred Points of Good Husbandry*.
47. *FARMERS' WEEKLY* (1985) 6 September.
48. LEWIS, M. R. (1984) *Fish Farming in the United Kingdom*. Department of Agricultural Economics and Management, University of Reading.
49. LEWIS, M. R. (1979) *Fish Farming in Great Britain*. Department of Agricultural Economics and Management, University of Reading.
50. GODDARD, J. S. (1981) 'Fish farming', in: HALLEY, R. J. (ed.) *The Agricultural Notebook*. Butterworths.
51. MINSTRY OF AGRICULTURE, FISHERIES AND FOOD (1988) *Survey of English Wine*. MAFF.
52. THELWALL, D. (1988) *Prospects for UK Mohair Production*. Report for British Angora Goat Society, and Food from Britain.

ANCILLARY RESOURCES

53. WATKINS, C. AND WINTER, M. (1988) *Superb Conversions: Farm Diversification: the farm building experience*. Council for the Protection of Rural England.
54. HOWICK, E. M. (1989) *Conversions of Redundant Farm Buildings: The interactions between farmers, planners and market forces*. BSc Honours Thesis, Seale Hayne College.
55. MINISTRY OF AGRICULTURE, FISHERIES AND FOOD/DEPARTMENT OF ENVIRONMENT (1989) *Planning Permission and the Farmer*. HMSO.
56. MINISTRY OF AGRICULTURE, FISHERIES AND FOOD (1986) *Agroforestry: a discussion of research and development requirements*. MAFF.
57. EVANS, J. (1984) *The Silviculture of Broadleaved Woodland*. Forestry Commission.
58. INSLEY, H. (ed.) (1988) *Farm Woodland Planning*. Forestry Commission Bulletin No. 80, HMSO.
59. BLYTH, J. *et al.* (1987) *Farm Woodland Management*. Farming Press.
60. CRABTREE, J. R. (1989) *Reorientation of UK Forestry Policy*. Paper presented to Agricultural Economics Society, Aberystwyth.
61. HEILIGMANN, R. B. *et al.* (1986) *Marketing Timber from Private Woodland*. Ohio State University.

Chapter 5

HOW SHOULD THE FARMER RESPOND?

Generalisations about policy change, marketing needs and alternative possibilities inevitably mask the tremendous diversity existing in the farming industry. The challenge to the individual farmer is to translate these observations into the context of *his/her* farm, and review the strengths and weaknesses of the bundle of resources that must be utilised to generate a livelihood. There may be a case for many farmers to continue as food-producing specialists. Others are likely to respond in different ways to the pressures for change. Those farmers who establish alternatives must fully review prospects before committing themselves to what may be major business changes. The methods of carrying out this appraisal form the subject of this chapter.

ASSESSING THE DEMAND

The first consideration with any alternative product from farmland must be whether or not there is a market for the product. It is not sufficient for the farmer to respond to generalised statements of growing demand. The pathways between producer and consumer are not always defined clearly and a much greater concern with marketing is likely to be necessary than with known products for known markets.

Market research should be conducted at an early stage. It can be carried out by farmers themselves or by professional market researchers. Market research can be conducted by the farmer when the product contemplated is likely to have a local market. Where markets are more dispersed, it is wise to make use of professional researchers. It is easily within the capabilities of a farmer to approach retailers or hoteliers with samples of a product. It is

151

even possible for farmers to conduct simple questionnaire surveys without recourse to the services of a consultant. Educational institutions with an involvement with marketing might be prepared to conduct case studies on behalf of individuals or at least direct interested farmers to texts that might make self-conducted market research a real possibility. However, a badly worded questionnaire is worse than no questionnaire at all. Nevertheless, surveys can yield information about consumer interest in, say, door-to-door vegetables or producer retailed milk.

Market research can also be conducted on established products. It is desirable that it should be. The demand for a product can change as a result of changes in taste, increased competition or many other factors. Producers in farming have tended to ignore such changes. The need for reviewing markets for alternative products is perhaps greater than in the case of an agricultural product where the responsibility for market research can be more readily foisted on policy makers or agencies like the Milk Marketing Board. This does not mean that there is no room for market research at an individual firm level with conventional products, merely that there may be less of a necessity to act individually.

Whether the farmer is contemplating an additional alternative or reviewing the market for an established alternative, there can be no substitute for structured market research. It is important to differentiate between products satisfying local demand and products satisfying a wider demand. But in both cases an overview of the market is desirable. Taking the case of the farmer contemplating yogurt production, what must he do? Firstly, he must find out generalised trends in yogurt consumption. These can be readily obtained from Milk Marketing Board sources (see Figure 5.1). But yogurt is a differentiated product. There are different fruit yogurts, Greek-style yogurts and a number of similar products like *fromage frais*. We know little about the markets for these, but it is highly likely that there will be some evidence in food marketing journals which could be unearthed. Talking to specialist wholesalers or to retailers may yield further information. All of this tells us about demand here and now but the real interest must be about demand in the future. Furthermore, it tells us nothing of potential competitors. Is there a niche market or could large-scale producers push small-scale producers into oblivion? Against whom would the producer be competing in local markets? In a rapidly changing market place there is no room for complacency. A heightened awareness of the market place is essential.

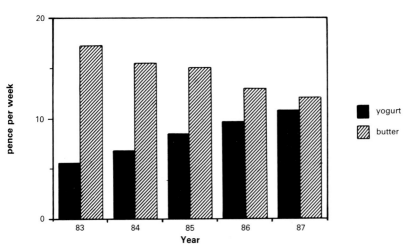

Figure 5.1 Weekly spending on butter and yogurt

Recognition of the importance of market research is given in the new Diversification Grants Packages. Although, in a classic cart-before-the-horse approach, the Diversification Grants preceded the introduction of Feasibility and Marketing Grants, these new grants should be welcomed. Grants of 50 per cent of the cost of feasibility studies are given with a ceiling of £3000 grant for individuals and £10,000 for groups. The Feasibility Grants are, of course, payable only for eligible enterprises (see Table 5.1). Small-business consult-

Table 5.1 Eligible enterprises for Feasibility, Marketing and Capital Grants under the Farm Diversification Scheme

Processing farm produce or timber
Craft manufacturing
Repair and renovation of agricultural machinery
Farm shops and PYO
Holiday accommodation
Catering enterprises
Facilities for sport/recreation
Educational provision
Livery
Facilities for pony hire (LFAs only)
Buildings to let (if the use is one of the above)

Source: ADAS (1988)

ants or local enterprise agencies may be able to help with market research. The National Farmers' Union Marketing and Business Services provide useful information on a range of alternatives. ADAS has limited expertise in market research on alternatives, although it grant-aids marketing studies.

It should not be implied that market research can act as a substitute for entrepreneurial flair. It can complement it, but it is no substitute. There must be a source for the ideas that are scrutinised by market research. High levels of entrepreneurial flair may yield windfall profits to the producer. The techniques of market research can readily point others in the direction of the proven successes but they are likely to be less adept at breaking new ground.

REVIEWING THE AVAILABLE RESOURCES

The successful establishment or management of alternative enterprises will hinge around the farmer's ability to reconcile the demand for a product with the effective deployment of his available resources to create it. The farm's resources can be divided into the classical trinity of land, labour and capital. The strengths and weaknesses of each in relation to alternative enterprises must be considered.

LAND

Land as a resource is made up of a number of attributes. Physical and biological attributes exert a strong influence on the productive potential of land. In addition, location is an attribute that must be recognised. In the past, location exerted an important influence on agricultural land use. In the future, and particularly with changing patterns of consumer behaviour, location could again assume its former importance, especially when farmers are marketing alternative products.

Farmers are familiar with the biological and physical attributes of land in an agricultural context. They are aware of the range of conventional agricultural crops that can be grown and the potential of the land under grass and grazing livestock. This information is institutionalised in the Agricultural Land Classification produced by MAFF, which purports to offer a semi-objective consideration of land quality. However, the quality of the land in relation to an unconventional crop may not be known. The uncertainty surrounding the climatic limits of certain exotic crops like evening primrose should lead to considerable caution by growers. Furthermore, success in one season may not be a sufficient basis for a significant expansion of acreage in subsequent years.

Low-grade agricultural land may not be low-grade land in an alternative use. Steep lowland valley slopes in southwest England or the Welsh Borders may be amongst the most productive silvicultural sites in the country for hardwoods or softwoods. The difficulties associated with agricultural utilisation mean that the land is likely to be designated low-grade. The grading system, however, applies only to its agricultural value and cannot be used as a reliable indication of its value in alternative uses. Poorly drained land may have potential for fish ponds for either farming or recreational use. Patches of land with limited agricultural value can be developed to enhance significantly the value of a shoot. There is a danger that a blinkered assessment of quality from an agricultural perspective leads to a failure to review the suitability of alternatives.

The spirit of agricultural fundamentalism lurks close to the surface when agricultural advisers and farmers assess land quality with respect to alternatives. In spite of the change of heart reflected in *Development Involving Agricultural Land*,[1] those involved in agriculture are rarely willing to see good-quality land taken for an alternative, even if the alternative use could substantially outyield conventional products. Instead of considering the optimum package of land uses on a given area of land, the alternatives are often irrationally condemned to areas of poor-quality land.

Historically, location exerted an important influence on agricultural land use patterns. Market gardens and milk producers occupied land close to markets because of the perishability of their products. As transport costs have declined and the speed of transport has increased, so agricultural land use patterns in the UK have come to be determined more by the quality of the land than by proximity to the market place. The pricing arrangements of the Milk Marketing Boards have further reduced the significance of location in the case of milk production. Consequently, location in respect to markets has ceased to be a significant consideration in most routine agricultural production.

In the case of alternative enterprises, location can be a vital consideration, primarily because of accessibility to markets but also with respect to other factors. Pick-your-own enterprises require not only proximity to markets but also a frontage on to a road with significant amounts of passing traffic. Although the physical and biological characteristics of land are likely also to be a relevant consideration, location in relation to the road system and population distribution are matters of primary importance. There will be many farms with high potential for horticultural production in the Fens with very low potential for PYO sales. Wherever products are sold from the farm to final consumers, or where the producer delivers a product direct to consumers, location will assert itself as a significant factor in enterprise success. A poor location may be temporarily sustained by product differentiation or effective promotion but it is a burden that will intensify if and when competition increases.

Farmland close to built-up areas is both burdened and benefited by its location. An urban fringe location presents particular hazards of uncontrolled access, dumping and theft. Survey work has indicated that large numbers of farmers can be affected. Some of these adverse factors can operate on alternative enterprises, too. However, the scope for alternatives which take account of, and benefit from, such a location cannot be ignored. The value of grass lets in urban fringe areas is frequently high because of the demand for grazing horses and ponies. The introduction of set-aside and the associated grants may create a charter for 'horseyculture' in the urban fringe, where marginal cereals areas can be turned to alternative uses. Further acres set aside on headlands offer the scope for privatised bridleways.

Buildings may also command higher values for alternative use in urban fringe settings as long as planning permission can be obtained. The enhanced capital value of land with development potential can turn a smallholder into a multi-millionnaire if he is prepared to sell. Not much land is beyond monetary value to those that farm it.

Location can also be important in the case of tourist and recreation enterprises. Tourist enterprises can be established either at tourist destinations or on tourist routes. The location of a farm may be a major feature especially where farms overlook attractive landscapes. The uniqueness of particular vistas or accessibility to recreational attractions may increase the potential of certain locations. Farms should not be seen in isolation from the surrounding area and the district's strengths and weaknesses as a location must be considered. There is a potential problem for farmers at the end of no-through roads where casual trade is siphoned off before it ever reaches the farm at the end (which may, incidentally, be in the most attractive setting).

As well as being relevant in respect of population distribution, accessibility and surrounding amenity, location of various features within the farm may also be important. Where woodland is peripheral to a farm it is less easily exploited as part of a game enterprise.

Where buildings, which are ideal for conversion to holiday flats, look out on a slurry lagoon, the vista and the odour may leave something to be desired. Furthermore, outlying buildings which have ceased to be agriculturally useful may be too distant from electricity, for example, to justify the additional costs of equipping them with services. Thus, location at a general and a specific level can be both a constraint or a factor giving significant advantages to certain farmers.

LABOUR

The composition of the farm workforce and the labour requirements of the farm should be considered when alternative enterprises are being contemplated or developed. Different types of alternative enterprise create very different demands on the workforce. Some are labour intensive and create a demand for labour at unsocial hours. Others require specialist skills which may not be present in the existing workforce. These demands on labour should be identified at the outset.

In many farming systems there are significant peaks and troughs in the demand for labour at different times of year. The conventional way in which these peaks and troughs have been examined, at least on livestock and mixed farms, is by means of labour profiles. These can be constructed using standard data for labour requirements of different types of enterprise which can be found in any standard book on farm management. Two simplified examples are given in Figure 5.2 to show characteristic labour profiles for arable and livestock farms. Arable farms are characterised by high labour demands in spring, late summer and early autumn. There are major troughs in winter and around midsummer. Livestock farms, especially dairy farms, create a much more even labour profile with a peak at silage and/or haymaking.

The labour profile of different types of farming can be considered alongside the labour demands of any alternative enterprise. Lowland, non-dairy, livestock farms are likely to have a summer trough in their labour demand once forage conservation has been completed. This trough may coincide approximately with peak labour demands from alternatives relating to tourism. In the uplands, delayed forage conservation may lead to competition for labour with overlapping periods of peak demand. A similar situation of competition may exist in the arable east. Furthermore, the peak labour demand for other types of alternative such as PYO is likely

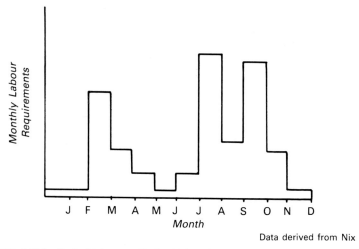

Data derived from Nix

Figure 5.2(a)   Typical labour profile for 50 ha wholly arable farm

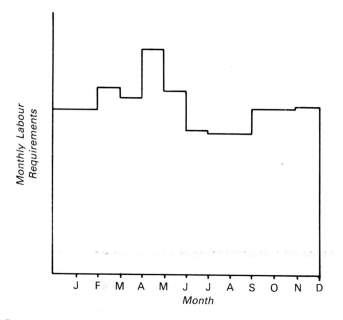

N.B. not on same scale as 5.2(a)   Data derived from Nix

Figure 5.2(b)   Typical labour profile for 50 ha wholly dairy farm

to be around harvest time for field crops. In this case it is much more difficult to mesh together the troughs of conventional enterprise labour demands with the peaks of the alternative. It should be noted that the labour profile may be an inappropriate method of analysing labour demand on arable farms where gangs of workers are required for particular tasks. Accordingly, gang work charts should be substituted for labour profiles.

Labour which is fully paid for but under-utilised can possibly be channelled into alternative enterprises. The individually constructed labour profile is likely to be much more complex than the simple illustrated examples and slack labour may already be used effectively. Peak labour demands can be met by casual labour or contracting and, in reality, the troughs are rather more shallow than might be expected. The troughs should not, however, be assumed away. The farm woodland on an arable farm may be managed in the winter months, given that the requisite skills exist or can be acquired readily. Farmers and farm staff may contribute significantly to the farm tourist enterprise, even though in many cases it may be in a subsidiary role.

Certain types of alternative enterprise are likely to be associated with female labour, particularly tourist and catering enterprises. Historically, farmers' wives have been involved in certain aspects of mixed farming, especially in dairy and poultry work. The specialisation that has occurred in the farming industry has led to a decline in the demand for female labour. If, as seems reasonable, the farm *family* is seen to comprise the workforce, it is likely that female labour in the family will be under-employed. Given the tendency of the farming community to ascribe different areas of work to different sexes it can be difficult to absorb this labour in conventional farming. It is scarcely surprising that many farmers' wives have developed tourist enterprises as an alternative to the small enterprises that were traditionally important in providing them with supplementary income.

Alternative enterprises may create a need for additional skills, and training in these skills is not well established in the UK. The expertise required may range from social skills relating to dealing with the public on the farm, to marketing and selling associated with value added enterprises, to technical know-how associated with certain new farm products such as horticultural crops or trout. It is not sufficient to justify alternative enterprises on the basis of slack labour. The qualities of the workforce must also be considered, particularly its skills and flexibility. The possibility of additional training being found to develop any new skills should also be explored.

The provision of training and education relating to alternatives is less formalised than is the case with agriculture, and in many cases will not be available within the agricultural college set-up. Few agricultural colleges offer education and training in food technology for those interested in on-farm processing. Courses relating to leisure and tourism will often be put on in technical colleges rather than agricultural establishments. In many cases those developing or contemplating alternatives need to acquire new skills, but the wide array of skills required and the differences in initial levels of knowledge make formalised training very difficult. As a result, a one-to-one relationship with a small business adviser may offer a more suitable opportunity for learning.

It may be entirely reasonable to establish alternative enterprises with additional labour. Where specialist skills are required or where there are highly seasonal demands, it may be difficult to meet labour requirements internally. Demand for part-time labour in tourist enterprises will frequently be met outside the farm household. The low level of economically active females in rural areas is likely to provide a pool of labour which will often willingly accept casual work. At a time when the farming industry is shedding labour, any additional employment opportunities are likely to be received favourably by rural communities. In addition, planners may look favourably upon developments that create jobs in areas of relatively high unemployment.

CAPITAL

The capital resources of any business can be considered in terms of how available they are for redeployment without threatening the financial integrity of that business. There is little on a farm in terms of fixed assets which cannot be cashed in should business rationalisation require it. Land or buildings can be sold to raise capital. There is no necessity for farmers to own all the machinery that they require. Contractors can provide a substitute for most types of mechanical work. If capital is not available on the farm it may be sought from normal sources of lending. With certain types of alternative enterprises, grant aid must be recognised as a potential source of capital.

Alternative enterprise opportunities may be constrained by capital limitations. The very reason that alternative enterprises are

being sought earnestly by many farmers is the financial precarious-
ness of their agricultural businesses. Such farmers may be unable
to find the funds to develop capital intensive alternative enterprises.
The contrast between the capital required for purpose-built self-
catering chalets and the establishment of a farmhouse bed and
breakfast enterprise is an example of the differences between two
types of tourist enterprise. Similar significant contrasts in capital
requirements will exist between the development of coarse fisheries
in existing water surfaces and their conversion to trout farms.

When contemplating alternative enterprises it is a prerequisite
that the opportunity cost of capital is considered. If the capital
available can be located more profitably in conventional farming or
in a building society, that is its appropriate placement. However,
where unused buildings or neglected woodlands exist, the absence
of any current contribution should not lead to continued neglect.
It may be possible to reactivate what is effectively idle capital. Such
resources can be developed to generate an income stream where
nothing was being produced before. The resource stock of the farm
must be considered from a total business perspective rather than
a farming perspective, and this applies to capital as much as to
land.

Many new schemes have been developed to promote small businesses and to stimulate the 'enterprise culture'. It is clear from
ministerial speeches in the late 1980s that the agricultural sector is
expected to behave more entrepreneurially as government subsidies are reduced. Those diversifying should contemplate entering
schemes such as the Enterprise Allowance Scheme or the Business
Expansion Scheme. Few conventional agricultural activities are
likely to attract the interest of outside investors. A number of
unconventional enterprises might attract outside funding. Advice
should be sought before opting in to any schemes but there are
these additional sources of capital which might be especially valuable to the new entrant with a sound business idea.

SEEKING HELP

Farmers involved with, or interested in, alternative enterprises will
frequently have to look further afield for advice and help than when
they are dealing with conventional products. The combined forces
of MAFF and ADAS have provided a great deal of support for
farmers, but in many cases there may be inadequate expertise
within ADAS to deal with assistance for some types of alternative
enterprise. The recent introduction by ADAS of standardised

appraisal forms for assessing the scope for alternative enterprises ensures that all those in the field are using the same comprehensive *aide-mémoire*. The introduction of various grants relating to diversification makes the ADAS officer a natural contact point. Often, however, farmers will be forced to look beyond their conventional contacts in order to get advice and financial support. The introduction of charging by ADAS for almost all of its services is likely to further diminish its attractiveness.

Organisations that might be of assistance to farmers concerned with alternative enterprises can be found in the public sector (e.g. the Forestry Commission), in the private sector (e.g. Fountain Foresty) or in the voluntary sector (e.g. Woodland Trust). In addition to hunting down the possible organisations, farmers must learn to discriminate between the different sources of advice being given about subjects which may be relatively unfamiliar to them. It is also important that farmers can separate out those organisations which *must* be consulted from those which *might* be. Appendix II lists a number of organisations which may be able to offer the farmer advice and grant aid.

There are essentially two types of public sector agencies. Firstly, those agencies which have a particular interest in one type of alternative enterprise – the Forestry Commission and the tourist boards are agencies of this type. Secondly, there are multifunctional public agencies like local authorities which are responsible for a wide range of functions from planning control to weights and measures and food hygiene. Frequently, local authorities are seen as negative agencies that constrain and obstruct and yet they can also be enabling agencies aiding the development of alternative enterprises in various ways.

There are a number of agencies which must be consulted for certain types of alternatives. Many developments will require planning permission. Tree-felling operations on more than a small scale will require felling licences and trees may also be covered by preservation orders. A failure by farmers to keep to statutory procedures can easily lead to ill feeling developing between officials and farmers. The tree felling of one Kent farmer in the early 1980s damaged the public image of farmers generally, and became a *cause célèbre* for conservationists. It benefits no one to denigrate intervention that is necessary by law or to ignore necessary official procedures. Thus, the first priority of farmers with an interest in alternative enterprises is to consult with those agencies that must be consulted.

In addition to those organisations in the public sector that *must* be consulted there are those that *can* be. Some county councils have developed woodland advisory and tourism advisory services. Furthermore, there may be a strong case for voluntary consultation with planners prior to the preparation and submission of a planning application. Minor modifications might be recommended by planners which would render an application more likely to succeed and avoid the expense of additional applications and the frustration and cost of delay or appeal.

There are many voluntary agencies that might be called upon to give advice. These range from producers' associations such as the Farm Shops and Pick Your Own Association, to the Goat Farmers' Association and organisations like the Woodland Trust. These can act as advice agencies and disseminators of information, and in the case of the Woodland Trust will actually plant and manage woodlands on behalf of farmers.

One voluntary agency which is showing increased interest in diversifying the rural economy is Business in the Community, the parent of the Local Enterprise Agencies (LEAs), which originated in urban areas but have moved out into the countryside. Local Enterprise Agencies tend to be staffed by business persons on secondment from industries other than agriculture but some have shown a desire to reach out and involve the farming community. The East Devon Small Industries Group (an LEA) carried out a major survey of redundant farm buildings. The wider range of free advice and their ability to act as signposting agencies to other sources of advice, make the LEAs valuable ports of call in the search for information and ideas on diversification. It numbers many farmers among its clients.

The private sector organisations offer a third source of advice. In some areas of activity – such as woodland management and farm tourism marketing – they are reasonably well developed. In other cases – such as farm museum developments – it may be much more difficult to track down private sector expertise. Such organisations offer advice for sale, rather than the warm glow of satisfaction that sometimes suffices in the voluntary sector. At present many new consultancies are being established. However, the interested farmer should beware of those who become experts in alternatives overnight and carefully scrutinise their credentials. The particular management needs of alternative enterprises differentiate them from conventional farm enterprises in a variety of ways. Farm business consultancies may not have the requisite skills to deal with, for example, visitor-based enterprises. There may be a case for using small-business consultants who are likely to be more market oriented than farming related consultancies.

Some of the agencies that provide advice also provide grant aid. The Tourist Board grant aided many developments until the recent cessation of the grants. The Rural Development Commission (and its predecessor CoSIRA) have funded many barn conversions for industrial uses (including tourism). MAFF has recently entered the fray with its diversification grants. Where funds are obtained from public sector bodies, farmers will frequently have to forego a certain amount of independence of decision making. At times the procedures that must be followed to obtain grant aid may seem bureaucratic and unnecessary. They may *be* bureaucratic and unnecessary. These procedures, though, are the price that must be paid in order to obtain the assistance. The assistance must be balanced against any inflexibility created for the farmer, but it may be sufficiently important to make the difference between a profitable and an unprofitable alternative.

Anyone contemplating an alternative enterprise of a particular type is likely to consider approaching established practitioners. At times these established individuals may willingly impart advice and ideas. Others, fearing local competition, may be less forthcoming, and their action is entirely understandable. To share a limited market with a competitor may be to threaten the survival of the business. However, it is not impossible for the potential entrant into the farm tourist industry to visit a variety of farm tourist establishments in other parts of the country in an effort to isolate the ingredients in success. Where local competition is not an issue the information divulged may be different, and more pertinent.

The challenge to the farmer is to make effective and discriminating use of the organisations that purport to offer useful advice. A commitment to a single source of advice, especially where the farmer's knowledge of the alternative is weak, is hazardous. It may be equally hazardous to dither interminably so that by the time an alternative enterprise is established, the windfall profits that can be obtained by early entrants to successful alternatives can be missed.

APPRAISING THE PROSPECTS

There are no universally applicable techniques which can be used to determine whether or not a particular alternative enterprise is a viable proposition. However, it is reasonable to assert that no alternatives should be established in the absence of financial appraisal. The appropriate technique may be influenced by such factors as the nature of the alternative, the financial position of the farm and whether or not the appraisal is likely to be used to attract outside finance or grant aid.

There is a danger that the techniques of appraisal used may reflect a continued production orientation of the farmer. It is by no means sufficient to confidently predict building conversion costs and subsequent occupancy rates, or yields of goat's milk. What matters is the relationship between costs and returns and marketing strategies may have a significant effect on the price at which units of the product are sold. The techniques of appraisal must be sufficiently wide-ranging to explore the markets for the product, the qualities of the product, where it will be produced and sold and how it will be promoted. All the elements of the marketing mix (see Chapter 3) must be incorporated into the financial appraisal.

The type of alternative enterprise being considered should influence the choice of appraisal method. Where enterprises are small and scarcely impinge on other aspects of the business, it may be possible to 'test the water', having carried out only rudimentary financial appraisals. Where considerable commitments of capital are required, as in purpose-built chalets, more elaborate and thorough appraisal techniques should be used. It is always important to recognise the extent to which alternatives can affect existing farm enterprises. The creation of a large trekking enterprise on a hill farm is likely to require a thorough reappraisal of the whole farm.

Many working farmers shun financial appraisal techniques in spite of the existence of many textbooks detailing a range of methods of analysis.[2-4] This irrational prejudice against analytical techniques and an over-reliance on intuition may hinder the prospects of businesses in the challenging years ahead. Whether the individual farmer carries out the appraisal or not, it can still be argued that an understanding of the principles of financial appraisal is worth while. It must be stressed that intuitive appraisals are often a poor second best.

In the past, farmers who diversified usually did so without the benefit of significant amounts of written information. Often they had to rely on estimates published in one of the standard farm management pocketbooks.[5,6] The contemporary situation is very different. Not only is there a great deal of coverage of diversification in the farming press but there are many publications to choose from. The big banks have begun to produce glossy brochures on diversification, perhaps appealing more to the city-slicker-turned-farmer than the impoverished farmer. The former is more likely to be able to pay the very high real rates of interest. Two publications attempt to outline the returns to different types of alternatives. [7,8] If these are treated with the appropriate caution, the figures offered constitute a starting point for a more individually based analysis. But in the case of an alternative, today's booming product could readily become tomorrow's oversupplied industry. There is a premium on up-to-date information backed up by an ability to forecast the medium-term future.

GROSS MARGINS PLANNING AND RELATED TECHNIQUES

Gross margins planning has been widely used by farmers and agricultural advisers. Frequently, gross margin figures are incorporated in more detailed planning techniques. Sometimes they are used to give a general indication of the contribution of different enterprises within a business. It is always important to bear in mind their limitations for high gross margins can be confused with high profitability.

A *gross margin* is a partial measure of enterprise performance calculated by subtracting variable costs from an enterprise's gross output. *Variable costs* are considered to have two distinguishing qualities. Firstly, they vary directly with the size of the enterprise, e.g. sprays for cereals. Secondly, they must be entirely allocatable to a particular enterprise, e.g. casual labour for potato picking. The *gross output* is the total value of all output from the enterprise. The gross margin concept is widely understood and provides an estimate which can readily be compared with widely published standards. However, even with conventional farm enterprises, what is a variable cost can differ from farm to farm. Contracting costs and casual labour are variable costs; machinery costs and full-time labour are fixed costs. Either factor could be employed to carry out the same task on different farms.

In the case of alternative, particularly tourist, enterprises, the concept of gross margin is often replaced by the concept of *margin over direct costs*. Within alternatives, it may be relatively easy to allocate certain items of cost which, in a normal farming enterprise, would not be allocated. The margin over direct costs is the gross output of an enterprise less all costs that can be directly allocated to it. It would be highly misleading to compare a margin over direct cost with a gross margin. In the case of a bed and breakfast enterprise, 'repairs, renewals and maintenance' and 'fuel and electricity' appear as allocatable costs, when for farm enterprises these would be pooled as fixed costs (see Table 5.2). Both these measures are appropriate but they should never be subjected to direct comparisons. Where this muddling of margins occurs because of different conventions or different sources of advice, it is vitally important that the differences are understood.

Table 5.2 Gross margins and margins over direct costs compared

Gross output	£100	£100	Gross output
Variable costs	Food Casual labour Advertising Laundry Others	Food Casual labour Advertising Laundry Full-time labour Fuel/electricity	Direct costs
Gross margin	£70	Repairs/ renewals Maintenance Others	
		£50	Margin over direct costs

The profitability of the two bed and breakfast enterprises is the same in this example. All that differs is the costs that have been excluded from the gross margin calculation.

If the primary purpose of a gross margin calculation is a comparison with standards, the gross margin or margin over direct costs will remain a tool of limited value in alternative enterprise appraisal. Gross margins for many alternatives are likely to be far more variable than is the case for conventional products. Not only are they subject to variation because of weather, but location is likely to exert a powerful influence on the profitability of many alternatives. A margin per bedspace per annum should be much higher on a Cotswold farm than on a Caithness croft, in that the former is favoured by a much longer tourist season.

Where a single enterprise is being established, the figure estimated for margin over direct costs may be of more value than a conventional gross margin in that a greater proportion of the fixed costs are overtly incorporated into the analysis. Although the concept has been most widely employed in the analysis of tourist enterprises, it is equally applicable on any alternative enterprise which is relatively independent of traditional agricultural enterprises. It could be used in value added enterprises and would take into account fixed items which would be excluded from gross margins. Where a lumpy input will be used over several years it is more appropriate to include an annual charge than a once-off lump sum which would distort figures for comparative analysis from year to year.

Gross margins and margins over direct costs are thus both useful calculations in financial appraisal. They can be used as indicators of enterprise performance and can constitute integral parts of other forms of analysis.

BUDGETING

Budgeting is widely recommended as an aid to decision making in farm businesses and the bundle of budgeting techniques is equally applicable as an aid to decision making about alternative enterprises. Partial budgets are used where recommended changes do not affect the whole farm. Where larger scale changes are being contemplated, whole farm budgets can be used to appraise the prospects. There is a whole family of techniques which come under the general term of 'budgeting'.

Partial budgets

Where alternative enterprises have impacts on certain other parts of the business activity of the farmer, but not sufficient impact so as to require a review of the financial situation of the farm as a whole, partial budgeting is recommended. A partial budget looks only at the costs and returns that are affected by a proposed change. The costs and returns of the alternative will be considered alongside the revenue foregone and costs saved as a result of making the proposed change (see Figure 5.3).

The decision to develop an alternative enterprise should only be made when the benefits on the right-hand side of Figure 5.3 exceed the costs of the left-hand side by what the manager deems to be an appropriate margin. It is most important that all the effects of

Debit	Credit
Additional Costs	Additional Revenue
Revenue Foregone	Costs Saved

Figure 5.3   A partial budget

Define objectives

↓

Categorise resources

↓

Estimate enterprise type and size

↓

Estimate physical inputs and outputs

↓

Estimate input costs and output revenue

↓

Estimate fixed costs

↓

Summarise options

Figure 5.4   The procedures involved in whole-farm budgeting

the proposed change are considered and that the riskiness of any proposed change is taken into account. Partial budgets might be especially useful where small quantities of land or buildings are taken out of agricultural use. Thus, a farmer taking 1 ha of wet land out of agricultural use for the creation of a fishing pond should look at the revenue foregone as well as the income gained. Building conversions for non-agricultural uses may create additional costs for livestock housing if they have been used for this purpose and replacement is necessary. Partial budgeting is a relatively straightforward technique which is highly pertinent to the type of appraisal problem generated by alternative enterprises which do not have implications for the whole farm.

Full- or whole-farm budgets

Where major business changes are being contemplated it may be desirable to analyse the business prospects in greater depth. Often such comprehensive appraisals are carried out when farms are sold or when new managers or owners take over a business. In the present state of flux in the farming industry, where individuals are contemplating large-scale adjustments (which may or may not include alternatives) there may be strong reasons for increases in the number of whole-farm appraisals.

The objective of the whole-farm budget is to explore the financial opportunities in depth. It is important that such an appraisal should be related to the objectives of the owner. Such budgets are likely to be most useful where they are carried out by the farmer rather than by an external agency.[9] The rudiments of the process are summarised in Figure 5.4.

The whole-farm budget can be built up either by looking at gross margin contributions or by looking at net margin contributions to overall profitability. The net margin approach is derived from the calculation of margin over direct costs which is used widely with alternative enterprises. The whole-farm budget looks at the net contribution of each proposed enterprise to overall farm profitability (see Figure 5.5).

Where the proposed developments are being implemented over a period of years it is necessary to explore the financial growth path of the business. Some types of alternative enterprise may not immediately generate their ultimate long-term contribution. Thus, where a building conversion for several tourist units is being implemented over a period of years the accommodation will not all generate contributions in the first year of trading. Furthermore, performance figures in early years may be poor because of the high initial marketing costs, without the possible benefit of return visits and personal recommendations. A development budget can be drawn up to show a snapshot of the business at future times.

If a development budget represents a financial snapshot, then a cashflow budget can be compared to a cine film. The cashflow budget itemises and summarises the flows of cash into and out of a business over relatively short periods of time (see Table 5.3).

Table 5.3 Schematic representation of cashflow budget

	Time period			
	January	February	March	etc.
(a) Cost/ expenses				
(b) Returns/ receipts				
(c) Balance for month (a–b)				
(d) Cumulative balance				

The manager should use the cashflow budget to identify the vulnerability of the business to months when returns are low and costs high. The bank manager may request such information to back up a request for borrowing and to set overdraft limits. Long-term

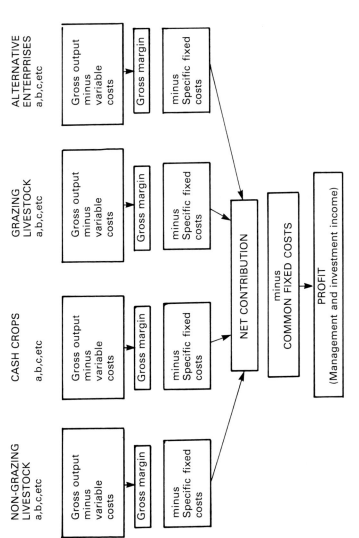

Figure 5.5 A framework for budgeting of enterprise contributions to farm profits

viability is an irrelevance if short-term cashflow crises cannot be survived. The seasonality of many alternative enterprises makes cashflow budgeting desirable in many cases and essential in some. The cashflow budget is likely to be used retrospectively as a benchmark against which to monitor performance (see Chapter 6).

CAPITAL APPRAISAL

Gross or net margins offer means of assessing the financial contributions of different enterprises to a business as a whole. Such figures are also likely to be used in whole-farm budgets where major reorganisations are desired. The financial ups and downs of a business are best explored by means of cashflow budgets which are valuable indicators of where a business is going. Sometimes the focus of attention is likely to be the capital invested in the business, or the capital that could be invested to improve business performance.

It has been argued that the various ratios used to assess the capital position of a business from the balance sheet have been oversold to the farming community. Thus, it can be seen as desirable for the farmer to maintain a reasonably high net worth (he should own at least 50 per cent of the assets tied up in his business) to avoid overtrading (the situation where short-term assets are exceeded by short-term liabilities) and to avoid too many of his assets being tied up in fixed assets, which can inhibit flexibility.

The assessment of a return on capital to a business is an indication of past performance. Frequently, the measure employed is the return on tenants' capital. This is the return on all capital except buildings and land. Owner occupiers may prefer to calculate a return on all capital.

The return on *additional* capital invested should be of major interest to the farmer concerned with alternative enterprises. The extent to which capital appraisal is required is likely to be influenced by the nature of any proposed investment. Where large sums of capital are invested – as in the case of chalets and building conversions – it may be vitally important to estimate the performance of capital tied up in the enterprise. If the development is based on borrowing, it is important to know that it can be serviced.

Various ratios are employed to assess the return on capital. Often the investor wants to know how long he must wait to pay back the investment. This is known as the *payback period* and is calculated by looking at:

$$\frac{\text{Total investment}}{\text{Annual profit of investment}}$$

An alternative measure is the return on capital which is measured in percentage terms:

$$\frac{\text{Profit}}{\text{Capital required}} \times \frac{100}{1}$$

Favoured enterprises will be those that produce short payback periods or high returns on capital.

There are also more elaborate methods of appraising capital investment. Where major projects are being undertaken or contemplated, and outside consultants are involved in aiding decision making, *discounted cashflow* may be used in investment appraisal. However, whilst the theoretical validity of the technique is not disputed, it is unlikely to form part of the farmer's repertoire of appraisal methods. Its value as a technique is likely to be greatest where there are major investments at a time which will produce returns over many years into the future. Forestry is a classic example of an enterprise where discounted cashflow is used, although short-term and medium-term tax avoidance has also been a powerful influence on forestry investment decisions.

RISK ANALYSIS

Whatever type of financial appraisal is being conducted, it is normally desirable to take into account the *riskiness* of an enterprise or an investment. Different individuals have different attitudes to risk: some are very risk-averse, whilst others will happily involve themselves in a business gamble. A strong case can be made for both groups to know about the riskiness associated with alternative enterprises.

For a number of reasons risk can be greater with alternative enterprises than is the case with many established enterprises. The main contributory factors can be summarised.

- Novel enterprises frequently produce more variable costs and returns than known enterprises because of the levels of knowledge and technical skill on the part of the operators (and sometimes advisers).
- There is rarely any market support for alternative products. No one will bale out the producer who offers a product that no one wants.

- There can be uncertainty as to what proportion of the product is sold at different prices, e.g. direct sale prices versus wholesale prices.
- The performance of many alternative enterprises is affected by location and the adverse or beneficial effects of certain locations have not been calibrated.
- The tastes of the public may be more fickle with alternatives than with conventional farm products. This point is especially important with non-essential, luxury products.
- Weather will be important to the performance of some alternatives.
- Alternatives require a change in the business orientation of farmers. The recognition of this need (Chapter 3) may precede the ability of farmers to reorientate themselves to the market place, by the application of appropriate marketing techniques.

The combined effect of these risk-inducing factors should not be to induce despair but to encourage a rational appraisal of risks. It is important to combine the appraisal of risk with the financial appraisal of the new enterprise.

The following analysis of risk focuses on the consideration of risks in gross margin and net margin calculations. The same principles of evaluation apply whatever system of financial appraisal is being conducted. The fundamental principle is that it is not sufficient to look at the single most likely outcome. Instead, a range of possible outcomes should be explored which reflect the uncertainty associated with the enterprise(s) under scrutiny.

This analysis can be conducted by producing a matrix of possible outcomes which reflects the likely range of costs and returns. The matrix should focus on the factors that are most variable and are major influences on costs and returns. These factors are often known but, if not, can readily be found by trial and error. It is useless to perform such analyses with cost and return figures plucked at random. They must have some validity to the particular situation under investigation.

Two examples will be used to illustrate the principle.

Example A   *Self-catering holiday enterprise*

It is recommended to carry out the analysis in at least two stages. The first stage establishes the principal contributory factors to gross output. These will be

- The occupancy rate (the number of weeks let).
- The average price per unit per week.

On the basis of investigations, an individual may find that the likely range in weeks let is 8–20 weeks and the average price per unit could be £80–160/week. The matrix of possible outcomes would be as shown in Table 5.4.

Table 5.4 Gross output matrix

		Occupancy (weeks let)		
		Low	Medium	High
		8	14	20
		(£)	(£)	(£)
Price per unit	Low £80	Worst 640	1120	1600
	Medium £120	960	Most likely 1680	2400
	High £160	1280	2240	Best 3200

The next stage is to consider cost variations. Low, medium and high gross output figures are transferred from Table 5.4. The range of direct costs likely to be met is estimated in the rows of the matrix. The resultant grid estimates the range of possible outcomes for the particular investment under scrutiny. When making a decision, the likelihood of occurrence of particular outcomes should be taken into account as well as the range (see Table 5.5).

Table 5.5 Net margin matrix

		Gross output		
		Low	Medium	High
		640	1680	3200
		(£)	(£)	(£)
Direct costs	Low £400	240	1280	2800
	Medium £600	0	1080	2600
	High £800	−160	880	2400

The net margin matrix does not include the initial investment cost. A further matrix could be produced which subtracts an annual charge for the capital from the net margins. The margins must be sufficient to justify investment and reward the farmer for his efforts.

Example B   *Firewood sales*

Returns from attempts to revitalise small woodlands on farms will
be highly variable. The principal influences on returns to a woodland management enterprise are the yield of timber and the price
received. The volume of standing timber in derelict woodland
cannot be established easily and the price received is likely to be
influenced by the degree of competition, the method of sales and
the presence or absence of trees with higher value than firewood.

Table 5.6 Gross output matrix

		Timber yield		
		125 (t/ha)	187.5 (t/ha)	250 (t/ha)
Average timber	8	1000	1500	2000
price (£/t)	16	2000	3000	4000
	24	3000	4500	6000

The costs of production in the woodland enterprise will vary
considerably. If there is slack labour and the farm has adequate
equipment to extract the timber, then the costs may be relatively
low. Where a contractor is brought in to fell the timber, production
costs will rise significantly. The individual farmer should consider
the range of costs he could encounter and produce a gross or net
margin based on these as a guide whether to go ahead or not.

The number of stages of analysis carried out using the sensitivity
matrix depends entirely on the nature of the alternative enterprise
under examination and the level of detail required by the farmer.
In some cases, it may be desirable to construct more than one
matrix in calculating gross margins. In a pony trekking establishment, gross output is likely to be dependent on three key variables
as follows.

● Price per trek.
● Frequency of treks (occupancy rate).
● Length of season.

The value of borage may be a function of yield, oil content and oil
price, and all three should be considered as important variables,
unless any variables are 'fixed' (such as oil price).

There may be some situations where it is considered that one 'box' in the matrix is not applicable. In the case of holiday chalets, it is unlikely that the farmer will achieve the longest season *and* the highest weekly rental, for the highest rental will only be achieved over the relatively short period of greatest demand. The 'box' in question can be deleted from the matrix and the next highest (or lowest) figures considered in subsequent calculations.

The matrix of possible outcomes is no more than a means of aiding a decision. It cannot make a decision for the farmer who must decide in the light of the probability of different outcomes and his own attitudes to risk. The accuracy of the method depends on the ability of the farmer to estimate accurately costs and returns. This ability will vary from enterprise to enterprise and location to location.

The consideration of risk is equally important in budgeting and capital appraisal. The objective is the same: to estimate the range of possible outcomes and their effects on business performance. Where farmers have access to computers, it may be possible to generate a range of outcomes very easily. The computer, however, can only speed up the calculations. It does not make decisions on which calculations to make or what figures to use as input.

The analysis of risk should not be stuck on to a conventional business appraisal technique as an afterthought. It should constitute an integral component of the analysis carried out to assess whether or not to go ahead with the establishment of an alternative enterprise. The analysis should take into account the production aspects (what yield at what cost?) and the marketing aspects (what price?); it is especially important to consider the prices that could be received in the case of alternatives, for the uncertainty of the market place must be given full consideration.

Where operators are confident about the prices relating to production but are less certain about the market prices of the product, the risk analysis can focus more directly on the price received for the final product. In a number of types of alternative enterprise there are significant premiums to be obtained from selling direct to the public. This is the case with deer and fish farming and with many value added enterprises, the products of which would otherwise pass through a wholesaler's hands. It is not difficult to construct a simple table indicating the returns to a particular product varying with the proportion sold direct and the proportion sold wholesale (see Table 5.7).

Table 5.7 Assessing the effect of marketing outlets on gross output for a given quantity of outputs

Direct sales (%)	100	80	60	40	20	0
Value of sales (£)	29768	23814	17861	11907	5954	0
Wholesale (%)	0	20	40	60	80	100
Value of sales (£)	0	3969	7938	11907	15876	19845
Gross value (£)	29768	27783	25799	23814	21830	19845

For example, 10 t trout: retail value £1.35/lb; wholesale value £0.90/lb

The lowest likely outcome in Table 5.7 should be equal to or greater than the breakeven point (the point at which revenue equals the costs). Even where a high volume of direct sales is predicted confidently, future competition can erode margins by stealing customers, especially where the initial enterprise is in a less than optimal location.

The initial appraisal of the prospects for alternative enterprises must go beyond the forms of analysis advocated for conventional agricultural products in two respects. Firstly, it is crucial to consider the risks of any particular alternative in considerable detail. Secondly, the appraisal should be seen as an integral part of a marketing plan which takes account of the market for the product and the means by which it is made available by the producer. This entails a thorough analysis of the business environment and the factors that can and cannot be controlled. It matters not whether the product is Christmas trees or farmhouse cheese, organic crops or pony plots. Marketing plans must be developed which interweave an appraisal of markets, an appraisal of resources and an analysis of the ability to produce alternative goods and services profitably.

REFERENCES

1. DEPARTMENT OF ENVIRONMENT (1987) *Development Involving Agricultural Land.* Circular 16/87. HMSO.
2. WARREN, M. F. (1982) *Financial Management for Farmers.* Hutchinson.
3. BARNARD, C. J. AND NIX, J. S. (1979) *Farm Planning and Control* (2nd edn). Cambridge University Press.
4. NORMAN, L. AND COOTE, R. B. (1971) *The Farm Business.* Longman.
5. NIX, J. S. (1988) *Farm Management Pocketbook* (19th edn.). Wye College, University of London.
6. SCOTTISH AGRICULTURAL COLLEGES (1985) *Farm Management Handbook.* SAC.

7. *FARM DEVELOPMENT REVIEW* (published monthly). Cambridge Publications.
8. PARKER, R. R. (1986) *Land: New Ways to Profit*. Country Landowners' Association.
9. GILES, A. K. AND STANSFIELD, J. M. (1980) *The Farmer as Manager*. Allen and Unwin.

Chapter 6

RUNNING ALTERNATIVE ENTERPRISES

When a new farm enterprise has been established successfully the management problems do not disappear, but they do change. Sound management requires effective *planning* (the subject of the previous chapter) and effective *control*. This chapter looks at how management control might be exercised on alternative enterprises. It addresses itself to two principal questions, as follows.

- Is the business performance adequate in relation to specified aims?
- What action must be taken to ensure that the enterprise performance remains adequate?

It does not matter if the alternative enterprise was established to gratify a personal whim or to satisfy a market demand. There is still a need to monitor the performance of the enterprise to see if it is meeting the aims established by the manager.

The performance of alternative enterprises must be considered in relation to other changes in the farm business and to change in the economy as a whole. The relative importance of alternatives may increase as the result of such factors as the imposition of quotas. This may lead to a respecification of aims and objectives for the alternative. The challenge of management control is to ensure that the business stays on course in spite of changes in the business and social environment. At such times as there is great turbulence in the wider environment the control function at an individual business level becomes more crucial.

There are many elements to management control. Performance must be monitored and the influences of controllable and uncontrollable factors on enterprise profitability taken into account. In the case of alternatives, the effect of changing levels of competition must be given particular attention. It is also important to explore the relationship between enterprise size and enterprise profitability, for there is some evidence that the conventional wisdom on size and efficiency does not apply. Changes in government policy, particularly in relation to taxation, can influence enterprise performance and must be considered. Finally, it is important to recognise that, as well as having a financial impact on the business as a whole, many alternative enterprises have a significant social impact on the farm household.

Operating an alternative enterprise is not necessarily more complex than running a conventional farm, but at times a different approach is required. In part, these differences relate to the need for a greater market orientation (see Chapter 3), but they also arise from the characteristics of alternatives and the difficulty of imposing external standards on performance.

Assessing Performance

Monitoring performance and dealing with unexpected outcomes is of the utmost importance to the farmer. In order to assess performance there must be reliable data on costs and returns and a yardstick against which performance can be measured. This can be obtained from external sources, as in comparative analysis, or internally derived where budgetary control techniques are employed. 'Comparative analysis' is the term employed where comparisons are made between a business and standard figures such as can be found in regional farm management surveys.

Performance can be assessed effectively only if an adequate pool of information exists. There are a number of reasons why data relating to alternative enterprises may not always be especially reliable. Alternatives frequently begin in a small way: tourist enterprises may owe their origins to doing a favour for overbooked friends; woodland enterprises may have begun as a result of the depredations of Dutch elm disease; PYO enterprises may have been established because of the absence of casual labour for picking fruit. The small scale and diverse origins of alternative enterprises are both likely to contribute to poor record keeping. As well as being small, many alternative enterprises have been characterised by being temporary. The once-off sale of dead elm is unlikely to stimulate formal record keeping. A final reason why formal records have not always been kept may be that alternative enterprises are often the domain of the farmer's wife or other members of the farm household. The proceeds of such enterprises are often regarded as something outside normal taxable farm income. Thus, a whole range of factors have contributed to poor record keeping in the past. As alternatives become more important, the case for adequate records strengthens accordingly.

There is much evidence to suggest that alternative enterprises will become more important in the future. The recent policy documents of the NFU and the CLA recognise this, as do MAFF officials and politicians. Over time, the threshold between inconsequential source of pocket money and business enterprise will be crossed by many operators of alternative enterprises. In addition, many new entrants will seek to establish alternatives for commercial reasons. Records will be vital for both established operators and new entrants. In their absence the assessment of performance becomes little more than an intuitive appraisal of how an enterprise is performing. It should also be noted that adequate record keeping is essential for taxation reasons.

Comparison of the performance of one business with an externally derived standard has been used widely in agricultural education and occasionally at farm level. The principle is simple: an enterprise or farm performance is compared with a benchmark. The most widespread application of this method is in the Farm Management Survey. These surveys are carried out by regional centres which produce regular reports on their general findings and occasional reports on specific subjects. The findings of all surveys in England and Wales are then summarised in an annual MAFF report. A similar procedure is operated by DAFS in Scotland. As long as the limitations of comparative analysis are recognised, they can offer initial benchmarks against which to compare performance of traditional agricultural enterprises.

The scope for comparative analysis is more limited in the case of alternatives. There are examples of detailed costings of certain types of enterprise, e.g. surveys of private forestry are carried out by those universities offering forestry courses. There has been a significant amount of investigative work on farm tourism and individuals can compare their figures with the averages obtained from surveys. However, the very high levels of variation in costs and returns (see Chapter 5) limit the value of comparative analysis in this sector. Once-off surveys of organic farms[1] or articles in the trade press on subjects like deer, goats, trout or PYO can indicate roughly how other enterprises are performing. Observations should be made of these sources of information and any major differences between the published figures and the individual operator's performance should be explained.

The collection and storage of records must be carried out giving due regard to the purposes for which the data are to be used. Records are a means to an end and not an end in themselves. It is desirable that there should be data on cashflow to record the ebb and flow of income and expenditure and data on profitability, the sum of which will allow an assessment of profit and loss over an annual period. In each case there should also be budgeted figures which can be compared with actual performance.

Cashflow analysis is important when setting up a project and remains an important means of analysing subsequent business performance. Cashflow analysis is a very sensitive means of analysing business performance, for cashflow can be considered on a weekly, monthly or quarterly basis. Normally in alternative enterprises or normal farm enterprises the interval of analysis will be 1 month. The cashflow is budgeted into the future on an annual basis. When an enterprise is being established, the cashflow budget may extend over a longer period, but with a functioning enterprise and in the absence of major changes, a projection forward for a year is the maximum required.

Cashflow analysis is only of any real value when a budget is meaningful and the analysis of deviations from budget is full and rigorous. The deviations should yield clues as to what, if any, remedial action should be taken. In many cases the principal purpose of cashflow budgeting will be to support pleas to bankers for lending. The banker will want to see the evidence that a business can climb out of spending troughs. As many alternative enterprises generate an uneven flow of costs and returns, cashflow budgets can be employed to fix a level of maximum overdraft.

There may be sound reasons why actual and budgeted figures should differ. In 1985 and 1986 anyone running an alternative enterprise which relied upon attracting people out of doors had to contend with particularly adverse weather. Tourist and recreation enterprises suffered, as did PYO establishments. However, it is also possible to experience a decline in performance which demands rather closer scrutiny of the business. A decline in casual trade to a farm bed and breakfast enterprise may be a function of increased competition or failure to submit a card to the local tourist office. In the case of a PYO establishment, a decline in performance may lie in low yields, poor-quality produce, increased wastage, poor advertising, increased competition or a host of other possible factors. The cashflow analysis should be sufficiently detailed to aid the isolation of the cause. If problems can be identified at an early date, appropriate remedial action is more likely to succeed. Where a cashflow situation is deteriorating, a long delay can seriously weaken the business. It is possible for this deterioration to occur without the farmer being fully aware of it. This is likely to be the case when a combination of factors conspire together to worsen the situation. A number of major items of cost can increase and yields and demand can be low. When this is the case the importance of a prompt managerial response cannot be overstressed.

Profitability of a business can be explored over a longer time period by the preparation of trading budgets. The budgeted profit and loss accounts provide an annual benchmark against which actual performance can be assessed. The previous year's performance is often included for comparative purposes. There is no single correct way of conducting this type of analysis. In some industries complete enterprise costing is favoured where both variable costs and apportioned fixed costs are allocated to each enterprise in order to assess the contribution of each to overall profitability.[2] However, on many farms the allocation of fixed costs to enterprises can be extremely arbitrary. This is especially so in the case of the small mixed farm. A framework for analysing gross margins and allocatable fixed costs was offered in Chapter 5 (see Figure 5.5). This recognises that there may be residual fixed costs which cannot be allocated, but apportions costs as far as possible.

In the assessment of the performance of alternative enterprises, gross margins analysis is rarely used. The concept of margin over direct costs is preferred. This constitutes a halfway station on the way to complete enterprise costing but avoids some of the flaws of complete costing. Alternatives are frequently less dependent on a common pool of fixed costs. Consequently, the net contribution or margin over direct costs can be used in budgeting. The advantage of looking at budget and actual net contribution is that it leaves a smaller pool of common fixed costs wherein the problem may lie. As a result, there is a greater chance that any weaknesses will be picked up and allocated to specific enterprises rather than be hidden in the murky pool of fixed costs.

The value of any form of analysis of budgets and actual performance is affected by the precision with which transfers from one enterprise to another can be costed. On occasions there will be transfers from conventional enterprises to alternatives, e.g. farm-produced hay is likely to be fed to ponies used for trekking. Home-produced meat and vegetables may be consumed by paying guests. These direct transfers must be considered. It is also desirable to take into account the cost of redeploying labour from one enterprise to another. Where productive work is lost because agricultural workers are required as beaters, the costs of the transfer should be taken into account. On some occasions, productive labour time may be lost. At others, there may be little loss, as when under-employed arable workers are employed on woodland enterprises in winter months.

It is of no great significance whether managers identify financial weaknesses through budgetary control or through comparative analysis. The matter of significance is that monitoring procedures have been devised and implemented. Monitoring is not, though, an end in itself: it is a means of identifying strengths and weaknesses that can be followed up by rational action in pursuit of improvement.

The manager of the alternative enterprise may wish to narrow down the monitoring procedures so as to allow attention to be focused on key variables. One method that has been advocated is to list all the component cost and return items and to estimate the financial effect of a 10 per cent improvement in any one variable.[3] When the appropriate items have been listed, the chance of achieving the 10 per cent improvement is assessed subjectively by assigning a 'p value' (a probability) ranging from 0 (zero chance of achiev-

Table 6.1 The key variables table for performance monitoring

	A Effect of 10% improvement (£)	B Estimated chance of achieving 10% improvement p = 0 − 1	C Value of 10% improvement A × B (£)	
Occupancy rate	500	0.6	300	2
Price per unit per night	500	0.2	100	4
Food costs	600	0.8	480	1
Hired labour	150	0.1	15	
Repairs/renewals	200	0.8	160	3
Fuel/electricity	120	0.1	12	
Marketing	80	0.2	16	
Laundry	120	0.2	24	

ing it) to 1 (100 per cent chance). Table 6.1 indicates how such an analysis might proceed in the case of a hypothetical farm tourist enterprise. The procedure is completed by multiplying the potential financial gain by the estimated chance of achieving it (column A multiplied by column B in Table 6.1).

The purpose of such an exercise is to focus attention on improving performance and to ensure that the improvements attempted have financial significance. In the example cited, the farmer has identified the major components of cost and return. Food costs, occupancy rates and repairs and renewals emerge as the categories with the greatest scope for an improvement of performance. The reason for the difference on the probabilities for price ($p = 0.2$) and length of season ($p = 0.6$) lies in the farmer's belief that extending the season is a possible means of generating additional trade, whereas raising price is likely to deter it. It is also considered possible to adjust food costs by more careful menu planning, bulk buying and using low-cost domestically produced food. When the full array has been produced, attention can be focused on achieving those improvements generating the greatest response.

The above-mentioned performance monitoring technique is not one that will necessarily lead to the optimum. It will help to identify improvements but will not guarantee that the elusive optimum can be attained. Examples of optimising techniques can be found in farm planning texts but these are normally the preserve of professional consultants.[4,5] However, the technique described has the major advantage of simplicity and should lead to the farmer's mind being focused on the right range of variables. In essence, it is doing no more than trying to eliminate any slack in existing systems of management.

The techniques identified to date do not make explicit reference to marketing, although adjustments in marketing may be implicit in some of the variables explored in the foregoing analysis. Market research (see Chapter 3) can and should be used to monitor performance. Market research is frequently only thought of at the inception of a project (if then), but it also has a vital role to play in monitoring. It has already been noted that farmers are somewhat remiss in their attitudes to market research (see Chapters 4 and 5). The various components of the marketing plan (the four Ps: Product, Place, Promotion and Price) must all be monitored continually. An established product may face competition that has arisen since the project was established; competitors' prices may be marginally lower and persuade customers to shift their allegiance; the promotional methods employed may be failing; or the outlets chosen for the sale of a product may be failing to shift it in spite of more general evidence of a demand for it. The monitoring process should cover all of the four Ps and is an integral part of marketing management.

Monitoring and control lie at the heart of the management process in any organisation. These actions form a central link in the chain that begins with the preparation of budgets and ends with effective remedial action being taken. It can be summarised as a four stage operation as follows.

- Prepare budget(s).
- Identify key factors affecting profitability.
- Explain deviations from budget.
- Take remedial action where possible.

It would be naive not to recognise that this ideal world of efficient budgetary control bears little resemblance to reality. However, budgetary control is likely to become more and more important in the struggle for survival as the cost–price squeeze continues to tighten its grip.

Efficient monitoring and control are especially important in alternative enterprises. The process demands a combination of more conventional farm business management techniques with ideas derived from market research. The absence of any widely accepted performance standards puts the onus of performance monitoring firmly in the hands of the farmer or manager.

COPING WITH COMPETITION

Success can breed problems. An indication that an enterprise is successful may precipitate local competition, especially if there are few barriers to entry. Full caravan sites are difficult to disguise. So, too, is a steady flow of traffic visiting a farmhouse recreation enterprise. Additional acres put down to an alternative crop indicate an enterprise worth thinking about. Success can be deduced in a variety of ways.

With many alternative enterprises there are few barriers to entry. It is relatively easy to enter into production in the case of bed and breakfast or PYO enterprises. With some types of alternatives there may be a demand for more specialist skills or more significant commitments of capital. Both factors will tend to deter competition and create certain barriers to entry.

Where markets for an alternative are local, the entry of an additional competitor can reduce the market share of the original operator. The demand for doorstep deliveries of milk is limited. A new entrant can create considerable competition on price and service and the market share of the initial operator may decline. This same process can occur with riding establishments, PYO farms and tourist enterprises. Where there is only a limited local market the new entrant is often acquiring his slice of the cake at the expense of a neighbour. Where products are marketed over a more extensive area the same problems do not arise.

Over a period of time, the efficiently managed and well located enterprises will tend to succeed. Technical efficiency will not suffice in many cases, for location is as much a factor in enterprise success as the technical skills of the farmer. Late entrants can observe the mistakes of the early entrants and then themselves set up, sometimes in a more advantageous location. The dog-eat-dog element of some alternatives can be dealt with more effectively if a market orientation is adopted. There are two ways in which managers of alternative enterprises can come to terms with growing numbers of competitors. Firstly, they can differentiate their product in some way. Secondly, it is possible to co-operate with fellow operators in the production of joint marketing strategies. Examples of both approaches can be found amongst alternative enterprises on farms.

Product differentiation can take many forms. The farm gate seller of venison may offer guided tours of the farm to attract increased custom. The PYO enterprise may develop a cafeteria or snack bar, a play area or a covered retail area to attract a broader clientele. Distinctive labelling may be used in an attempt to create loyalty to the products of a particular farm shop. In recreation enterprises, a conscious but unobtrusive concern with visitor welfare might create additional personal recommendation. Home-produced meat and vegetables or local specialities on the menu may be used to differentiate tourist products. It is even possible to differentiate a product like firewood by such means as wood type and dryness. At least some customers will be able to discern the difference between the burning qualities of wet alder and dry ashwood. Thus, in various ways the farmer can seek to differentiate his product and make it more appealing to the consumer.

The second means of coping with competition is to co-operate with others in the production of a collective marketing strategy for a product or similar bundle of products. The most obvious example of this exists in the case of tourist and recreation enterprises. The Farm Holiday Bureau acts as a central agency for farm holiday groups, and numbers have been expanding rapidly in recent years (see Chapter 4). In the case of recreation enterprises, there are many examples of group brochures supported by regional tourist boards and widely distributed at tourist information points, in tourist accommodation and by visitors going to any one of the listed attractions. In both tourist and recreation examples it is hoped that the production of a collective brochure will ensure better marketing and a greater number of total visitors.

There are other ways of working collectively to enhance profits. Farmers can co-operate in the marketing of their products. Organic farmers have established co-operatives, as have deer farmers. Where a number of small producers can combine their output to reach a more distant and lucrative market place, co-operatives are likely to succeed. Where there are significant local markets that yield higher prices, then the co-operative is a less attractive avenue for marketing. The development of a number of organisations promoting specialist foods, normally on a county basis, is evidence of the belief in group marketing ventures. In addition to the co-operative marketing, other aspects of the overall marketing process can be addressed by co-operatives, including promotional activities.

Quality control is normally considered in individual production units. It is also especially important in group marketing ventures. One bad product in a hamper of specialist foods can tarnish the image of all. The same is true of providers of farmhouse accommodation. Thus, it is vital that groups have the power to police the members offering the product and reject those products which do not meet the desired standard.

The choice of an appropriate strategy depends on the producer, the product and the clientele. The established tourist enterprise – almost wholly dependent on return bookings and personal recommendations and working at, or near, capacity – may obtain little benefit from a farmhouse holiday group. Indeed, the costs of membership are likely to be greater than the benefits derived. However, all operators acting alone should recall that products have a life cycle, that tastes change and that the business environment is dynamic rather than static. Contemporary successes may be tomorrow's failures.

Competition is an ever-present threat to alternative enterprises. Operators must learn how to live with it if they are to survive and prosper.

ENTERPRISE SIZE AND EFFICIENCY

Many alternative enterprises are established on a relatively small scale. Many remain small, whilst others grow. The present pressures on conventional farm enterprises may lead those with alternatives to seek to expand them. It is therefore important to explore

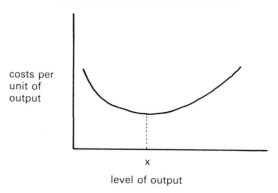

costs per
unit of
output

x

level of output

Figure 6.1   A typical average cost curve

the relationship between enterprise size and efficiency so that an efficient size of enterprise can be chosen.

A variety of factors has led economists to suggest that, up to a point, as enterprises grow, so they should become more efficient. In elementary economics, this relationship is explored by looking at the shape of cost curves (see Figure 6.1). The average cost curve shows the relationship between size and efficiency. In Figure 6.1, the most efficient level of output is at point x. There is some evidence that cost curves in conventional farming in the UK resemble this shape.[6] The shape is attributable to economies of size, which are considered to be greater in arable, than livestock, enterprises.

It is pertinent to ask whether the cost curves for alternative enterprises are likely to be the same shape as those for conventional products. More importantly, the factors influencing the shape need to be considered with special reference to alternatives. Two examples will be used to illustrate possible differences.

The first example relates to a farm bed and breakfast enterprise. It is assumed that there is spare capacity in the farmhouse and that the bulk of the furniture and furnishings will not need renewing. Consequently, the costs of establishment of a small-scale enterprise are likely to be very low. Sheets, towels, crockery and cutlery will be required, but little else. The average cost curve for this (see Figure 6.2) may bear little resemblance to that in Figure 6.1.

The lowest level of costs per unit of output will be achieved with a relatively small enterprise size. This is the result of being able to produce a small number of bedspaces at very low costs. Small may be beautiful for other reasons, too. Operators can avoid the need to comply with fire regulations and avoid rates and the necessity of

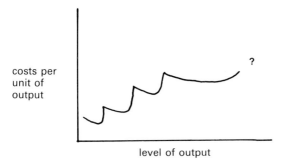

Figure 6.2   Hypothetical average cost curve for a bed and breakfast enterprise

planning permission. Furthermore, guests may choose farmhouses which take a relatively small number of guests out of preference for the personal attention they assume will be associated with small scale.

As the size of the enterprise increases so more costs may be encountered. The conversion of attic space may require new plumbing and wiring and, once the numbers of guests reaches the critical number of six, there is a need to comply with fire regulations which can necessitate considerable capital expenditure. When these financial hurdles have been crossed, it is possible that there are certain economies of scale. These might be derived from bulk buying or from savings on marketing. Alternatively, if the farmhouse accommodation becomes just another guesthouse for tourists and high occupancy rates cannot be achieved, costs per unit of output may well increase.

The second example relates to a deer farm. There is one major economy of size and this is the cost of fencing. Deer fencing is a major item of capital expenditure and the larger and more compact the holding, the lower the cost of fencing per livestock unit. However, set against this must be the very considerable premiums that can be derived from direct marketing of venison. If all the venison on a small unit can be sold direct, there are very real advantages to set against the disadvantage of higher fencing costs per unit of output. It might be asserted that an average cost curve for venison production would be relatively flat, although the precise shape will depend on the farmer's ability to obtain premium value for his output.

In some types of alternative enterprise there may be very real economies of scale. In woodland enterprises small size is likely to lead to relatively high costs per unit of output. Fencing costs are high as in the cost of deer farming but, in addition to this, there is the sophisticated and expensive machinery that characterises the contemporary forestry industry. To set against this particular example of economies of scale, it must be recognised that some woodland grant aid (particularly the British Woodland Grant Scheme) gives much higher levels of grant for small woodlands, or at least for small-scale working of larger woodlands.

It is important for each individual to establish the shape of the average cost curve. The reason is simple. Success at one level of output is no guarantee of success at a higher level. With certain enterprises, particularly those that can exploit slack resources with low or zero opportunity costs, the average cost curve may differ greatly from the classical U shape. There is a danger that enterprise growth will lead to a diminution of returns per unit of output, and, in some instances, a decline in overall profitability. Those who have established successful small-scale alternative enterprises should not be lulled into false optimism relating to expansion. An examination of the likely shape of the cost curve should temper decision making relating to changing scales of enterprises.

COPING WITH TAXATION

Various forms of taxation impinge on businesses in many different ways. The establishment and development of alternative enterprises on farms may alter tax liability and create different tax planning problems. The following observations are not a definitive statement on contemporary business taxation but identify the various categories of tax and some of the implications of alternative enterprises on tax. The taxes considered include the following.

- Income tax.
- Capital taxes.
- Value added tax.
- Rates and poll tax.

Farmers as a group have historically been treated relatively leniently and as special cases. Capital taxes were less onerous and capital allowances and the averaging of incomes created a benign fiscal climate. However, the advantages have been removed in a piecemeal manner and agricultural taxation is, in essence, no different from that of industries of an equivalent size, with the exception of rates relief on agricultural land and buildings.

Because agriculture has now been brought into line with the rest of industry, alternative enterprises are unlikely to have a major impact on income tax liability and avoidance. (The one exception is the case of commercial woodlands, which will be considered subsequently.) Inevitably, where alternatives are successful, thresholds may be crossed. When the joint income of a husband and wife exceeds a certain amount (*ca*. £27,000 in 1989) it is likely to be beneficial to be assessed separately rather than jointly. From 1990 onwards it is the intention of the government to tax married persons separately, thus removing certain anomalies. As many alternative enterprises – particularly those relating to tourism – lie largely under the managerial control of the farmer's wife, there is a clear case for separate assessment.

However, the moral case for separate assessment may be less important than the need to offset losses in one area of business activity against profits in another. There is little benefit in having alternative earnings as a lifeline if the rope is unnecessarily shortened by increased tax liability.

Commercial forestry enterprises were taken outside the income tax system in the 1988 Budget. The previous situation which allowed a generous tax shelter for high taxpayers has been abolished. The new situation puts all commercial woodland owners in identical situations with no relief on forestry inputs and no tax on outputs. There will no doubt be some furtive and judicious spending within agricultural enterprises which might have been more appropriately placed within a forest enterprise, but the Inland Revenue are fully aware of this possibility. There will undoubtedly be grey areas. Woodland that is not commercial will not be eligible for tax exemption and there is a need to define what is, and what is not, commercial woodland. Furthermore, there may be difficulties in deciding whether a silvicultural operation was for the benefit of trees, pheasants or amenity.

Capital taxes take the form of capital gains tax – which taxes certain capital gains on an individual's assets – and inheritance tax, which is a tax on property comprising an individual's estate immediately preceding his death.

Capital gains tax is assessed on chargeable gains on the disposal of assets after deducting allowable expenses. It is unlikely to be a major burden unless large capital gains are made and the gains are not then rolled over. The rollover relief allows the potentially taxable gains to be rolled into a new set of fixed assets. The tax is only due when the assets are finally disposed of. There are time limitations for rollover relief and there is scope for significant avoidance of capital gains tax in retirement relief. Capital gains tax is not payable on a main place of residence.

Inheritance tax is unlikely to be a burden to a farmer who indulges in sensible tax planning. There are many exemptions which make inheritance tax a tax on misfortune or ineptitude rather than capital. The 1986 Finance Act contains provision for tax-free lifetime transfers of wealth. If the transfers are made in the 7 years preceding death, a sliding scale of tax is applicable. Rates at death remain unchanged.

Since the mid 1970s, there has been an opportunity to obtain capital transfer tax relief on heritage land. This is most likely to be of relevance to larger landowners rather than working farmers, but the operation of the provisions merits attention, in view of the relationship to alternative enterprises. Land may be defined as heritage property if it is held to be of outstanding scenic, historic or scientific interest. Buildings, works of art and land can all be defined as heritage. In the case of land, relevant categories include Sites of Special Scientific Interest and landscapes of high merit. Reasonable steps must be taken to ensure public access to it. In the case of building maintenance, costs can be deducted and the property must be preserved. Heritage relief must be negotiated and is not something from which all can expect to benefit. It is, nonetheless, a very substantial benefit to those who are eligible. The potential relief which large estates can gain from manicuring their heritage assets may make the provision for visitors a small price to pay for keeping large estates intact.

Value added tax is a tax that is added to a product at every stage at which value is added. The standard rate of VAT is currently 15 per cent. Agricultural inputs can be either zero rated or standard rated. Where tax is paid, it can be reclaimed. It must therefore be in the interests of the farmer with alternative enterprises to retain the integrity of his business as a farm if there are no reasons to do otherwise.

Value added tax is not paid on businesses with a turnover of less than £23,600 in 1989. It may be in the interests of farmers to ensure that their non-agricultural alternatives can legitimately be considered as their wives' businesses. In the case of farm tourist businesses, there is a danger that exceeding the turnover limit could add 15 per cent to visitors' bills and deter custom.

Rates are local property taxes from which agricultural land and buildings are wholly exempt. The system of rates is being partially replaced by a poll tax called a 'community charge', a tax on adult people. This was introduced in Scotland in 1989 and is due to be introduced into England and Wales in 1990. However, the poll tax deals only with housing and not business property. This is a grey area in the case of farm cottages used for tourist activity. A farmer with a small holiday cottage enterprise may find it preferable to be rated as a business if the letting value of the property is low, or as a private furnished residence if the letting is high. Case law is needed to clarify the situation.

The principal anomaly between farm enterprises and alternative enterprises remains in spite of the Minister of Agriculture's suggestion that farming is just another industry. Thus, tourist cottages, chalets or flats are rateable, as are industrial premises located in converted farm buildings. As the scale of enterprise increases, so there is a danger of an increasing rate burden. This is especially so where a house ceases to be mainly a residence and becomes primarily a place of business. This occurs when a farmhouse becomes a guesthouse. It should be noted that the agricultural exemption from rates specifically excludes 'land kept or preserved mainly or exclusively for the purposes of sport or recreation'.

A further area where tax liability should be scrutinised carefully concerns national insurance. National insurance is now a less onerous tax on low-paid persons enabling the £43 per week (1989) threshold to be crossed without major disadvantage. This would make it easier to employ part-time staff earning, say, £60 per week, for national insurance would only be paid on £60 less £43, not on the total £60.

A final opportunity for tax avoidance in alternatives lies in the Business Expansion Schemes which allow very considerable fiscal advantages to outside investors, who can invest up to £40,000 per year in equity investments in unquoted companies and receive full tax relief from income and capital gains as long as the investment remains for 5 years.

It is not possible to do more than outline the main areas where taxation affects farmers and point out how it may affect alternative enterprises. It is most important to know the potential taxation pitfalls, and to be aware of any ways in which alternative enterprise expansion might change tax liability. As can be seen, there are some types of alternative enterprises – such as commercial woodlands – which are less onerously taxed than farming, whilst in other cases additional taxes, such as rates, are unavoidable, and certain agricultural benefits, such as income averaging, may be lost.

There are many references available on taxation which tend to focus either on conventional farm taxation problems or on taxation in a very general sense. Such texts can provide valuable information.[7] Generally, however, these tax manuals should be background reading to prime the farmer for a meeting with an expert, and should not be regarded as sufficient bases for making decisions. Taxation law is constantly changing and the cost of errors normally makes professional help a necessity and not a luxury. This is especially so where alternative enterprises begin to make a significant contribution to what was formerly a traditional farm business.

COPING WITH SOCIAL PRESSURES

Running an alternative enterprise can create social pressures. This does not necessarily differentiate alternatives from conventional farm enterprises but there are likely to be differences in the pressures created. This section identifies three areas in which there is potential for social pressures as follows.

- Pressures of unsocial hours.
- Pressures of visitor demands.
- Pressures of work versus play.

The demand for work at unsocial hours is common to many alternatives. It is not something that is likely to be a new experience to the average member of a farm household. Nonetheless, it is pertinent to recognise the unsocial hours that must be worked if the enterprise is to develop.

Whether the alternative is a tourist, or a PYO, enterprise, the peak labour demands are at times that many would regard as unsocial. The experience of getting young children to school at the same time as feeding farmhouse guests can create difficulties. With PYO enterprises the demand for labour is likely to be concentrated at weekends or in holiday times. Those who run such enterprises might also welcome free time in pleasant summer weather. Unless their enterprises are unsuccessful they are unlikely to get it. Furthermore, in the case of alternatives which rely on visitors coming to the farm, there is a need for someone to be present even if he/she cannot be employed productively.

The whims of the customers are experienced far more directly by the managers of many alternative enterprises. Many of the alternatives that offer a service rely upon the ability of the operator to identify demands of a social nature and respond to them. The services that surround a simple product like farm tourism augment the product and enhance (hopefully) the visitors' experiences. Citations to those who have won awards for serviced tourist accommodation nearly always mention the way in which the host approaches the visitors' needs. These additional services may include the operator spending time finding out about the visitors' interests and directing them to attractions that will appeal to their tastes or offering a free babysitting service while the visitors sample local pubs.

Visitors to farms will frequently not be from farming backgrounds, although such survey evidence as exists indicates that most have an interest in the countryside. There are likely to be some overlaps and some differences in attitudes and values. These may surface if there is considerable face-to-face contact between operator and customer. Social skills of tact and tolerance and an ability to direct behaviour without being in any way offensive are essential when dealing with visitors. Whilst a recalcitrant animal can be verbally and physically forced to act in a certain way, similar methods cannot be used on tourists or PYO shoppers.

Farms that are visited, for whatever reason, are the shop window of farming. The operators of these enterprises are the unofficial public relations team for the agricultural industry. Sometimes their work may be more effective than that of the official public relations experts. Operators of alternatives are in a position to influence public opinion. Sometimes guides and information panels in farm recreation centres and museums adopt highly defensive postures in response to criticisms made of farming by the wider public. Naive propagandist efforts are unlikely to persuade dissenting voices and may further exacerbate tensions between town and country. Farmers will be judged more effectively by their actions.

The way in which information is made available to the visiting public can be a major influence on visitor enjoyment. The art of interpreting the countryside in a way attractive to visitors must be acquired by the operator (or bought in). There are some outstanding examples of bad interpretation at farm museums and visitor centres as well as some examples of good interpretation. The cartoon style representations of pig, poultry and beef enterprises at Lightwater Valley in Yorkshire are a good example of how serious points can be made in an amusing, visitor-attractive way. The provision of a number of self-guided trails around an Exmoor hill farm, backed up by well-produced leaflets, constitutes a different but equally effective means of interpreting the environment to the visitor.

Where the public are invited on to a farm it is important to give due regard to visitor safety. Farms are dangerous places, especially to those unfamiliar with them. It is essential to be covered by adequate insurance and professional advice should be taken on insurance needs. However, a concern for visitor welfare should go beyond obtaining adequate insurance cover. It may be possible to zone visitors away from areas of agricultural activity that create dangers without diluting the farm experience that they may have come there to obtain. Leaflets on farm safety can be obtained from the Health and Safety Executive and can be displayed in an appropriate manner.

The origins of many alternative enterprises lie in the particular interests of a member of the farm family. This is entirely understandable. There is, though, a danger that alternative enterprises are disguised hobbies. Hobbies can be costly. If alternative enterprises originate in this way and it is intended that they make a financial contribution to the business, a line must be drawn between work and play. Game enterprises, farm museums, rare breeds enterprises or even tourist enterprises can all be manifestations of self-indulgence rather than business activity. The rarity of a piece of farm machinery is unlikely to appeal to the average visitor but may be the pride of an enthusiast's collection. A relatively recent machine actually working is more likely to appeal than an ancient machine of great rarity that cannot work.

If the operator can clearly establish his own objectives then there may be no reason why such self-indulgence should be criticised. The objective may be to collect exhibits as an investment rather than to generate an income stream from visitors. Alternatively, the objective may be to manage the enterprise at a pleasurable loss in order to limit tax liability. A difficulty only arises where the alternative is intended to generate income (and this is increasingly the case) but is being managed *as if* it were a hobby.

The sum of the social pressures created by alternatives can be considerable. Domestic routine may be disrupted severely and those who are unable to accept this disruption should exclude the most disruptive enterprises from their catalogue of possibilities. Since alternative enterprises frequently place demands on whole families, there is a strong case for collective decision making. It should also be noted that the ability to indulge in alternatives may be influenced by life-cycle considerations. There is clear evidence of adjustment in conventional agricultural enterprises and in responses to changes in family situations.[8] It is highly probable that where alternative enterprises impinge on the domestic economy, they too will be influenced by life-cycle influences.

The various ways in which alternative enterprise affects the social welfare of the farm household are relevant whether an enterprise is being established or expanded. Different individuals and households have different abilities. A degree of self-awareness of the social effects of alternatives is a desirable attribute for anyone involved in their management.

In conclusion, the monitoring and control of alternative enterprises raises a challenge. There are similarities between the monitoring and control operations that should be practised in any farm business. There are also certain differences. The evaluation of alternatives is not as highly developed a science as the evaluation of conventional farm enterprises. There is a greater onus on the farmer to establish appropriate procedures and targets. Certain difficulties can arise if the growth path of alternative enterprises is not considered in detail. It is important to avoid the trap of increasing scale and diminishing returns. Taxation must be taken into consideration when alternatives are being developed. Intuition should normally be backed up by professional advice. Finally, the social aspects of running alternatives must be recognised and the farm household must be capable of making the necessary adjustments and accepting the disruption that alternatives can create.

Perhaps the most important point for anyone involved in monitoring and controlling an alternative enterprise is to realise that its continued success depends on an awareness of consumer demand. This may change as a result of social changes or an increase in competition. The monitoring must extend to the wider patterns of demand for the product. Signals must be picked up and appropriate responses made. Only then can the operator maximise the likelihood of enterprise success.

References

1. VINE, A. AND BATEMAN, D. (1981) *Organic Farming Systems in England and Wales.* University College of Wales.
2. BARNARD, C. J. AND NIX, J. S. (1979) *Farm Planning and Control* (2nd edn), p. 537ff. Cambridge University Press.
3. ESSLEMONT, R. J. (1985) 'Waging war on waste', in: ROYAL AGRICULTURAL SOCIETY OF ENGLAND (eds) *Planning a Future with Milk Quotas.* RASE.
4. BARNARD, C. J. AND NIX, J. S. (1979) op. cit.
5. GILES, A. K. AND STANSFIELD, J. M. (1980) *The Farmer as Manager.* Allen and Unwin.
6. BRITTON, D. K. AND HILL, B. (1975) *Size and Efficiency in Farming.* Saxon House.
7. APSION, G. (1984) *Tax Saving Ideas for Farmers and Country Landowners.* Farm Tax and Finance Publications.

8. NALSON, J. S. (1968) *The Mobility of Farm Families.* Manchester University Press.

Chapter 7

ADAPTING TO A CHANGING WORLD

The farming industry, globally and nationally, has faced booms and slumps, periods of prosperity and periods of painful adjustment. The agricultural sectors of advanced Western economies have experienced different degrees of protection at different times from the vagaries of world markets. Whilst Japanese farmers are coco-oned in massive subsidies, those of New Zealand have been strug-gling after the removal of the relatively low levels of protection they experienced in the early 1980s. In North America, policy changes and major droughts have conspired together to reduce farm incomes. In Western Europe the early, cautious attempts at reform have been followed by more stringent price-cutting policies that will create adjustment challenges to the farming community. These adjustment needs have not been precipitated by local factors but by international processes. The productive capacity of the West's agricultural sectors has outstripped demand changes. The structure of the farming industry has meant that millions of individual decisions to increase output to counter the cost-price squeeze have simply exacerbated the problem of oversupply. These problems have been further compounded by a history of protectionist agricultural policy, especially in Western Europe.

The imperatives for change are global. The rural sector of developed Western economies is experiencing its own *perestroika*, a restructuring of major significance. The local detail of the restructuring varies from country to country and region to region. The countries of northern Europe have a rather more heightened awareness of adverse environmental change than their southern European colleagues. Consequently, the northern European initiatives in response to escalating agricultural support costs have reflected these environmental concerns, from controls over slurry application in the Netherlands, to nitrate controls in Denmark, to the promotion of Environmentally Sensitive Areas in the UK. Prevailing political ideologies are likely to colour the processes of change. Those in New Zealand and the US in the early 1980s had profound impacts on the course of adjustment.

Some are of the opinion that the problems of oversupply are temporary aberrations. A mildly tempered enthusiasm in the farming community greets news of failed harvests and global environmental disasters. Perhaps there will again be food shortages. Global cereals stocks are low in 1989 and another poor cereals harvest in the US could precipitate a short-term, but significant, crisis. The gainers would be farmers, the losers the poor of the Third World or the USSR, who are normally the beneficiaries of the surplus cereals swilling around the world markets. Should such an event occur, and it is by no means impossible, diversification issues would be brushed aside temporarily, only to reassert themselves a few years later.

Thus, the problems faced by British farmers in the 1980s are by no means unique. Their forebears faced similar problems 100 years ago when the Great Depression ended an earlier golden age. In the interwar years, the oversupply of agricultural products again produced considerable stress in the rural sector. Now there is a similar crisis of confidence. The causes of the present crisis will not disappear overnight. The early hopes that the problems would be shortlived have been proved false. Farmers cannot behave like latter-day Canutes when the tide of change is so strong. The onus is on the farming community, globally and nationally, to adapt to a changing world.

The economic problems of the farming industry continue unabated. Farm incomes in the UK were at their lowest level in 1989 since the Second World War. Although the incomes to dairy farmers have stabilised, those of all other sectors of the farming community remain threatened. Cereals producers are the second major group of farmers to experience winds of change. Their golden years of postwar prosperity have come rapidly to an end. The situation varies somewhat from region to region and farm type to farm type but the overall picture gives little hope for more than a temporary upturn.

The political climate in both the UK and Europe must be understood. In Europe, change is resisted and the power of the agricultural vote ensures a slow pace of reform. In the UK, the political climate has moved further against the farming community in the late 1980s. The wooing of the environmentalist is a far more important concern than easing the adjustment problems of the farming community: 200,000 farmers scattered around the shires are not the force they once were. Knighthoods have been awarded for exposing the flaws in agricultural policy and questions are asked as to whether ministers of agriculture, health or environment are appropriate persons to speak about agricultural concerns.

Environmental concern shows no signs of diminishing. The extent to which even the popular press is latching on to environmental issues is a sign of the appeal of environment to a broader cross-section of the community.

Some of the heat has gone out of the debate that waged after the publication of *The Theft of the Countryside*[1] in 1980 but, nearly a decade on, few doubt the potency of the issue. Marion Shoard's latest book[2] is more scholarly and less polemical but focuses again on the demands of the public for access to a rural resource that is threatened in various ways. While *The Theft of the Countryside* coincided with the heated debate about the Wildlife and Country-side Act, *This Land is Our Land* has coincided with a review of Countryside Commission policies relating to access. Access provides an issue which is likely to retain its importance into the next decade.

Food and health issues remain prominent. The almost hysterical response to the *Salmonella* and *Listeria* issue in 1988/89 indicates the extent to which public interest can be raised. The dietary debate of the mid 1980s has begun to affect the demand for food. Many of the dietary changes of recent years have reversed a number of long-standing trends. The consumption of red meat is declining. Brown bread has grown in demand, as have margarines high in polyunsaturated fatty acids. It has recently been asserted that:

> There is little doubt that factors other than economies are having an impact on food choice. The growing consensus of opinion is that consumer attitudes are increasingly affecting individuals' food consumption habits. *The desire for quality and an increasing awareness of health issues are becoming important issues in food selection.*[3]

Elsewhere it is noted that there are signs of convergence of dietary behaviour in many developed countries. The reports on diet and health in the early 1980s have been reinforced by governments, media and the retail sector. Food consumption patterns are influenced strongly by dietary concern, but the search for quality is increasingly evident. Whilst this might depress the producer of bulk animal products, it might hearten the producer of specialist foods, especially those which have a healthful image. The evidence of changing patterns of demand for food is overwhelming. The food retailers are not ignoring it. The food producers cannot afford to ignore it, either.

THE CHANGING POLICIES OF THE AGENCIES

The various agencies which produce policies that affect the way in which rural land is managed and the way in which the rural economy works have changed their policies to a greater or lesser degree. Diversification can develop in response to positive policy instruments encouraging alternative enterprises or in response to negative policies which have reduced incomes and encouraged diversification as a survival strategy. It is these various policies interacting in combination with market forces that create the business environment within which individuals and households must devise adjustment strategies.

At a European level, agricultural policies have begun to change. Spurred on by the growing costs of surplus, the precipitate response of milk quotas has been followed by a more complex bundle of policies. The Green Paper of 1985 recognised the failure of the CAP to balance supply and demand and the problems raised by the CAP in relation to environmental, economic and social policy. It was, however, far from innovative in its prescription, dismissing alternatives for three reasons as follows.

- There are not always adequate advisory services.
- Marketing structures are often weakly developed.
- The distorting effect of agricultural policy favours other products.

All these factors represent constraints, but none of them is unchangeable. It is an observation of supine stupidity thus to condemn alternatives, for the causes of all of the factors were rooted in the operation of the CAP, which the document sought to change. Instead, the preferred option was to look at new outlets for surplus products, e.g. the conversion of raw materials into bioethanol. There has been considerable interest in the application of biotechnology to agriculture and in the creation of new genetic material and new methods of production and processing. The logic is usually to look for products which, in European or national terms, are not in surplus.[4] The attempts to 'upgrade' sunflowers to render them suited to the UK climate seems to be a classic example of this misconceived logic. The principle of comparative advantage is ignored and soon world markets will be flooded by sunflower products, often from marginal locations where production should never have been established. The lessons from cereals should have been salutary. Unfortunately, they have been forgotten.

Whilst the technocrats are still learning, the Eurocrats have moved a little further. In *The Future of Rural Society*[5] we are informed that:

> The Commission's approach to rural development is guided by three fundamental considerations: economic and social cohesion, in a Community of very pronounced regional diversity; the unavoidable adjustment of farming in Europe to actual circumstances on the markets and the implications of this adjustment not only for farmers and farmworkers but also for the rural economy in general; the protection of the environment and the conservation of the Community's natural assets.

European policy is changing. Increased attention is being given to the *guidance* elements of the budget and less is being given to the *guarantee* element. Agricultural adjustment has emerged from the shadows. A review of structures policies in 1985 ushered in a change of emphasis, since when a number of measures have been developed.[6]

There is still little clarity in the way the EC wishes to aid the adjustment process. The Mansholt philosophy of the late 1960s was to get rid of marginal farmers and improve the efficiency of farming by increasing farm size. This resulted in the structures policies of the early 1970s. The desire to reorganise agricultural structures remains but it has been supplemented by objectives 'to help develop the social fabric of rural areas, to safeguard the environment, to preserve the countryside'.[7] Within this context, the interest in farm diversification and multiple job holding has grown, recognising finally the growing importance of opportunities outside conventional farming, on and off the farm.

Policies that enhance the prospects of the agro-industrial complex still smell of the production orientation that has created so many of the contemporary problems of the CAP. The new interests in rural communities represent a break from the production orientation. Whether it is a break with a recent tradition, back to a more distant peasant tradition, or a leap forward into a post industrial countryside, remains unclear.

The design needs of policies for agriculture that embrace wider rural community needs are likely to be very different in Greece and the UK. It is difficult to see how European-wide schemes can span the differences that exist. Thus, a partial renationalising of agricultural policy within broad guidelines set by the EC seems to be the only sensible way forward. A diversity of policy instruments may be necessary and desirable. Local problems can often be solved better by 'bottom-up' strategies rather than by the imposition of 'top-down' policies. If Eurocrats and politicians have the courage to act thus, the rural communities could be richer and more diverse as a result.

In the UK, MAFF has been showing signs of change. The more environmentally destructive grants have disappeared and environmentally friendly grants have appeared in their place. UK agricultural ministers took the lead in the development of policies relating to Environmentally Sensitive Areas. The recognition that there were public goods which merited public support was a major advance. Hitherto only in national parks and Sites of Special Scientific interest had there been compensation to farmers for providing public goods. The extension of this principle into ESAs also led to new procedures for compensation which avoided the complex negotiations of management agreements. However, the recognition of ESAs is also a recognition of the potentially damaging effects of intensive agricultural operations. Questions remain as to how much of the UK should be designated as ESAs. The early batch of designations was added to with unseemly haste, certainly before any evaluative methods had been sorted out or the limits to the public's willingness to pay for environmental goods ascertained.

Set-aside policies have also been introduced in spite of sound advice from the US that such policies were administratively expensive and economically illogical. Set-aside could have been used to satisfy wider environmental objectives. Instead of MAFF building environmental benefits into their set-aside scheme, it has been left to the Countryside Commission to launch a series of top-up payments in the summer of 1989 to provide the environmental bonus that set-aside could have given. This scheme is restricted to East Anglia but there are other parts of Britain, including areas in Scotland, that might benefit from similar schemes.

Set-aside has emerged as a scheme unlikely to be of any real significance to most amenity interests. In areas close to affluent populations, set-aside policies may provide a charter for 'horseyculture' by turning arable acres into pony pastures. The marginal gains from reduced cereals support hardly justifies the subsidisation of horseyculture.

Proposals to extensify agricultural production to reduce surpluses were shelved in 1987 whilst set-aside was implemented. In 1989 they are being reconsidered. They could contain elements promoting organic husbandry and may thus impact on diversification.

The initially lethargic response of ADAS has been replaced by a much more active concern in promoting interest in alternatives. There are, though, some gaps in its service, especially with regard to its knowledge about marketing. Nonetheless, its socio-economic advisers have been relabelled Rural Enterprise Advisers and are developing their skills in advising farmers on alternatives. The advisory publications on alternative enterprises are now no longer being produced. This is unfortunate, for such figures as they contained would have provided a benchmark against which to compare the figures of the proliferation of glossy presentations from banks and accountants.

The provision of the Diversification Grants package is the most direct indication of MAFF's commitment to alternative enterprises. By December 1988, 1300 farmers had taken advantage of the scheme, the majority to assist the development of leisure-related enterprises. The total investment remains small in terms of MAFF's total spending and, as the grants are part of a national scheme, they are vulnerable to public spending cuts. But they represent at least a token indication of support for alternative enterprises.

The winds of change are undoubtedly blowing through the corridors of MAFF. The Ministry of Agriculture has promoted rural restructuring in a number of ways. In 1987 the minister informed a conference that: 'there can be no doubt that considerable areas of the countryside will be looking for uses other than agriculture in the years ahead',[8] and firmly advised farmers at the Oxford Farming Conference to accept the need for off-farm earnings to supplement declining farm incomes.[9] However, what is most interesting in the incumbent government is how its predilection for market forces co-exists with an increasingly interventionist agricultural policy with respect to environmental issues. Quite how these contradictions in ministerial, and indeed in governmental, thinking will resolve themselves is unclear.

The agencies which represent other interests in the countryside must also be considered. The Department of Environment has become involved in the debate about the future use of rural land. Because of its responsibility for town and country planning, it is important that there should be effective interdepartmental liaison. Many alternatives require planning permission. Agriculturally stimulated change can thus have significant ramifications for other central government departments and there is a need for greater co-ordination of thinking than has sometimes been evident.

Some of the quangos that have grown out of the Department of the Environment have been vulnerable in an era of public expenditure cutbacks, when there has been a strong ideological commitment to market forces. However, the 'greening' of politics and consumers has ensured that these agencies will receive governmental benediction and funds into the foreseeable future. It may also be that their responsibilities will continue to grow, especially with regard to ensuring the quality of environmental public goods provided by such policies as ESAs. The debate about how and at what level farmers should be compensated for providing environmental goods looks set to continue. Some would argue that any type of price support should be conditional upon environmentally acceptable farming. Such a move would put the cereals barons of East Anglia under severe pressure.

Rather as MAFF has become a department of born-again environmentalists, the Development Commission and the Council for Small Industries in Rural Areas have become born-again enterprise agents, renamed the Rural Development Commission. Their commitment to grant aid in Rural Development Areas and their recent move into grant-aiding rural tourism make them a key agency in rural restructuring. One wonders how long the roles of RDC and the rural enterprise advisers of ADAS can be kept separate, and whether a Ministry of Rural Affairs would not make greater sense than the present situation with its divided responsibilities.

THE CHANGING POLICIES OF PRESSURE GROUPS

The National Farmers' Union, the Country Landowners' Association and the representative organisations in Scotland and Wales have been reappraising their policies in the light of the crisis. Both the NFU and the CLA produced documents within a few days of each other early in 1985 and both organisations have produced further reports which indicate the extent of the policy shift.

The agricultural and landowning organisations all recognise that rural land management must change. For the first time there is official recognition of the legitimacy of at least some of the criticisms. A very different view is offered from that of the heady years of the mid 1970s. Budgetary crises are recognised as realities. Environmental concern is acceptable and is no longer the preserve of eccentric 'ecofreaks'.

There are recurrent themes in the approaches of both the National Farmers' Union and the Country Landowners' Association. The laws of supply and demand cannot be flouted. A stronger

marketing orientation is desirable. Diversification is endorsed.
These and other changes represent a fundamental change from
earlier positions.

It is questionable how far these changes in thinking have permeated the ranks. Within the hierarchy of the NFU there remain
individuals attached to the traditional values and positions. The
shift required cannot take place overnight and there is bound to
be a degree of friction as the new policies filter down to the grass
roots. There can, however, be no doubt about the commitment of
the NFU to diversification; members of its marketing division have
researched alternatives and provide a highly marketing oriented
source of advice to members.

In addition to the activities of the major representative organisations, other bodies have shown growing interest in alternatives.
The Royal Agricultural Society of England has run many confer-
ences on alternatives and has broadened its concerns to the whole
rural economy. Enterprise agencies are moving into the countryside. The many organisations involved in different aspects of alternative agriculture have grown up, not always without a certain
amount of squabbling.

There have also been some significant changes in the attitudes
of the environmental pressure groups. The polemics of Marion
Shoard have been replaced by more temperate reports showing
how environmental concerns might be built into future agricultural
policies. The Council for National Parks, the Worldwide Fund for
Nature, the Council for the Protection of Rural England and the
Royal Society for the Protection of Birds have all tabled their
proposals, which try to build-in the public goods derived from
farming that were pushed aside by the CAP. Coupled with the
growing effectiveness of their lobbying, these reports are likely to
bear fruit. The ideas contained in them should be considered by a
forward looking farming community. The ability to act rationally
within a redefined set of ground rules is likely to be an important
factor in business survival in the future.

THE CHALLENGE OF ADJUSTMENT

The case for a reappraisal of agricultural policy is now widely
accepted. The range of policies advocated is wide and one can only
speculate about eventual outcomes. At a European level there is
an inevitable inertia resulting from the administrative procedures
but sudden changes are not impossible as milk quotas have proved.
At a national level the UK economy remains unstable and political,

economic and social changes are working against the farming community, creating a more volatile and less benign policy climate.

It is important to recognise the major social changes taking place in attitudes to work, leisure, eating and living. The countryside is a place associated with pleasant environments, leisure use, good food and a healthy existence. Many of these images have been subjected to a sustained challenge as people have moved into rural areas. Sometimes their expectations are to blame for their disillusion; sometimes the activities of the farming community may be responsible. These incomers and armchair policy makers for rural areas are often affluent and articulate. They can be the sternest opponents of coniferous afforestation or the strongest advocates of the locally produced farm product. It is important that they belong metaphorically to the latter group, seeing farmers as providers of goods and services they want, rather than seeing farmers as threats to the products, including scenery and wildlife, they currently consume.

Environmental issues matter increasingly to consumers. When the prospect of ecocatastrophe exists – as is the case with ozone destruction – the public begins to give environmental issues a higher profile. Politicians cynically follow. But over the last 30 years, environmental issues have grown in prominence. Those using the surface of the land must accept that their care must extend beyond narrow food-producing objectives. Tunnel vision produces problems like that relating to groundwater pollution. Such problems do not endear farmers to the rest of the community.

Farmers and agricultural policy makers must be aware of the wider implications of their actions and the wider policies impacting upon them. The absence of any clear vision of the future creates uncertainty and concern. Farmers know that the cost–price squeeze is likely to tighten its grip but have little idea of the business environment in which they might hope to be working in 5 years' time. In this respect, they are in much the same position as many other industries. The security of the last 40 years has rapidly evaporated. Business planning in a less certain world becomes a greater challenge.

There are a number of options that can be considered. The ability of dairy farmers to maintain their incomes in spite of quotas has been cited as evidence of slack that could be taken up. No doubt there is room for improvement on many farms. Farmers must eliminate inefficiency and pay much greater heed to the financial consequences of their actions. However, the amount of slack varies from farm to farm and some farmers must question seriously the wisdom of remaining in farming.

Although in the industry as a whole the ratio of capital to borrowing is sound, there is cause for concern. Land prices have begun to slip in some areas but have been held up in others by the amenity value of farms and buildings. This does not make profitable farming any easier. It simply makes the banks more willing to continue to lend, knowing that there is capital growth that could be retrieved if necessary.

There are significant numbers of farmers whose overdrafts have been rising. If overdrafts continue to rise as the price of land drops, the financial position of some could become precarious. The overborrowed tenant farmer is in an even worse position. There must come a time when it would pay some farmers to quit in order to cut their losses. Payments to outgoers have been contemplated by the European Commission and may provide the final incentive required.

Low input farming has also been advocated as a strategy for survival. However, the more heavily borrowed farmer on the more capital intensive farm is locked into a system from which escape is difficult. On other farms a process of de-intensification can be expected but while the policy rules fail to penalise unwanted products, de-intensification will be carried out by the visionaries rather than the pragmatists. The extensification measures currently being devised may provide the escape hatch sought.

These and other proffered strategies hinge on the assumption that there is a *farming* solution to this particular set of problems. For some, this is certainly the case. However, for two reasons it may be wise to look to non-farming solutions. Firstly, Ruth Gasson's work[10] has indicated the extent to which farmers and farm families pursue other gainful activities off the farm. Secondly, the demand for alternative uses of farmland may be increasing sufficiently to justify serious consideration of alternative enterprises.

Many farmers or members of farm households pursue careers outside farming at the same time as taking on certain responsibilities on the farm. Although in many cases these activities are entirely unrelated, it may be possible to expand them to compensate for the possible decline in farm incomes. The potential for using other gainful activities off the farm as a solution to declining household incomes will be influenced by the skills of those seeking work and the demands of the local economy and the extent to which policy makers can devise schemes for rural development which provide opportunities for household members who cannot find full-time employment on the farm. High levels of unemployment in some regions may make it difficult to find alternative work, but even in regions of high unemployment there may be possibilities for seasonal or part-time work.

The pursuit of the second option hinges on the recognition that farmed land, or the resources associated with farmed land, may contain opportunities for the establishment of alternative enterprises. In order to realise the scope for on-farm alternatives, farmers must first challenge the introspective production orientation that has permeated the industry. They must reorientate themselves to the market place and explore the demand for alternative products beyond the narrow range of policy-supported products that dominates the industry.

The recreational and tourist potential of farms has long been recognised and there is substantial evidence available to show the significant returns on investments in tourist enterprises. It has, however, always proved more difficult to make recreation enterprises profitable because so many recreational activities can be pursued in the British countryside without charge. The assumption that tourist enterprises on farms can only be developed in the traditional tourist regions has been proved wrong. Profitable farm tourist enterprises are not impossible in the lowland shire counties. Such data as are available point to highly variable returns to tourist and recreation enterprises, a fact that any potential entrant should be aware of.

'Value added' has become a new slogan in some circles and many must have heard the phrase without fully understanding what it meant. In essence, the principle of value added seeks to add as much value on the farm as possible. This can be achieved by alternative marketing, particularly direct marketing to the final consumer, or processing. Whatever method of value adding is employed it remains fundamentally important to recognise the difference between net value added and gross value added. It is the former that the farmer is seeking to achieve.

The search for new animal and crop products has gathered momentum and established alternatives are arousing increased interest. Frequently the producers of these alternative agricultural products endeavour to add value to them. These new products can be grouped into four main areas, although some of the products fall into more than one category. Firstly, the concern about health manifests itself in a demand for different methods of production (organic) and different types of food (low fat, etc.). Secondly, there is a demand for a variety of luxury foods such as venison, quail or expensive sheep's milk cheeses. Thirdly, the demand from ethnic minorities in the UK should attract interest, and there is some evidence of a response. Finally, craft interests may create a demand for products for basket making, weaving or thatching.

Areas of woodland or wetland and old farm buildings may have little potential for contributing to agricultural output, but these other components of the farm may have profitable alternative uses. The traditional neglect of farm woodlands has been replaced by concern and new support policies. Wetlands can be developed into recreation attractions or maintained for shooting or conservation. Farm buildings may have very considerable value as potential homes or workshops. The development of the earning power of these frequently neglected resources is increasingly in evidence.

Throughout the world, realisation is growing that agriculture cannot be separated out from the rest of the rural economy. Agricultural policies have historically looked at agriculture in isolation, as if the sector were not connected to many other elements in the economy. Narrowly defined policies centred on the expansion of production have generated many adverse consequences, often damaging the environmental products that an affluent population seeks. The restructuring of the rural economy needs a rural policy that recognises the centrality of agricultural production but takes heed of other demands for clean water, attractive landscape, leisure opportunities and wildlife conservation. As policy changes put price pressure on farmers, so it seems increasingly desirable to recognise their role in the production of environmental goods and reward them accordingly. If this requires the decoupling of price support from other aspects of agricultural policy, this is likely to benefit the consumer and the taxpayer. Furthermore, the wider support measures can be tuned more delicately to suit local conditions.

The successful development of alternative agricultural enterprises demands a much more rigorous understanding of marketing than has characterised farmers' thoughts and actions in the recent past. The marketing process is a way of thinking about the goods and services produced, not a way of getting rid of surpluses. It focuses attention on the consumers who are the ultimate arbiters of policies by means of the ballot box, taxes and pressure group activity. Farmers in pursuit of sectional interests have followed policies which may have yielded short-term dividends, but they have also created long-term problems. Consumers have not only reacted against many of the traditional products of farming but also against the way in which they are produced and the effects on the environment. The rational development of alternative enterprises is one way in which these communications failures between producer and consumer can be repaired. The marketing ideas and techniques that underpin their success can, and should, be applied to the whole range of farming activities.

One element that farmers must focus upon comprises the problems and potential created by location. Not all locations are equally endowed with scope for alternatives, particularly for those alternatives that involve direct marketing. As more alternative enterprises are established, so those in sub-optimal locations are likely to feel the pinch.

Policy support for alternative enterprises has developed to such an extent that many farmers are bemused by the options and uncertain as to when to take advantage of grants. Many hang back hoping that grant levels will rise. In doing so they may miss opportunities, such as the Farm Woodland Scheme, which in certain situations would seem to offer great scope for adding to the value of land. But many of the recent policy measures are less of a positive incentive into diversification and more of a negative incentive for continuing conventional farming practices. Policies for farm woodlands exist because of cereals support costs, not because of a genuine desire to plant more broadleaved trees. The bundle of policies is becoming more complex and it is incumbent on the farmer to be aware of the full array of policies.

The relative absence of policy support, particularly of market support via guaranteed prices, means that those who operate alternative enterprises must live in a harsher world of competition than their traditional colleagues. If you cannot sell your alternative product, it will not be given to the Russians at bargain basement prices while the government gives you a normal shop price. Furthermore, the heavy promotion of alternatives by agencies and others may cause significant changes in supply, as new entrants come in. In order for alternative enterprises to expand and thrive in this cut-throat environment, sound marketing judgement is likely to be more important than luck. Operators must be fully aware of changes in the business environment in which they are working.

The production aspects of some alternatives are not well understood. Individual farmers cannot normally be expected to research the production problems. At times, official support for research into alternatives on farms seems insufficient to meet the requirements of an industry that is struggling to come to terms with its deepest crisis for many decades. Whether penny-pinching politicians will provide the funding necessary is uncertain.

As well as being harangued by politicians, farmers are being lured into alternative enterprises by the glossy publications of banks and accountants. Their publications[11] might appeal to the yuppie farmer but they should be looked at with extreme caution by the rational entrepreneur. The use of a standard gross margin for a coarse fishing, or a pony trekking, enterprise is of extremely limited value.[12] But diversification is seen as a lucrative market. One cannot deny the organisations their right to sell their services but one can advise farmers to use them very cautiously.

The situation faced by farmers in the UK is not dissimilar to that confronting their colleagues in most other parts of the developed world. The extent to which alternatives have been introduced varies greatly within and between countries. New Zealand farmers have perhaps been the most effective diversifiers, this arising initially out of choice and, subsequently, out of necessity. In general, New World farmers seem less hidebound by traditions than their Old World counterparts. The more farmers have been exposed to market forces, the greater their propensity to diversify when conventional products have failed.

Alternative enterprises are being promoted by politicians with widely varying political views in Texas, in the UK and in New Zealand, all of whom view the diversification of farming enterprises as a crucial element in rural restructuring. The issue will remain with us.

Alternative enterprises are likely to be of increased importance to British farmers in the future. They are likely to be of greater importance throughout the developed Western world where over-production is rife and policy support costs high. The signals from the market place favour an exploration of alternatives. For many farmers there may be scope for the profitable establishment of enterprises rather different from the traditional enterprises on which they have come to depend. However, alternative enterprises do not offer an escape hatch for the agricultural industry as a whole. The opportunities for profitable alternatives are unevenly distributed. Whilst some may benefit substantially, others will be unable to derive any benefit at all. The challenge to the farming community is to identify these opportunities and develop and manage these enterprises with due recognition of the reality of the market place and the needs and desires of consumers.

REFERENCES

1. SHOARD, M. (1980) *The Theft of the Countryside*. Temple Smith.

2. SHOARD, M. (1987) *This Land is Our Land*. Paladin.
3. WHEELOCK, J. V. AND FRANK, J. D. (1989) *Food Consumption Patterns in Developed Countries*. Paper presented to Agricultural Economics Society Annual Conference.
4. COWEY, K. (1988) 'Biotechnology: the limits to development', in: WHITBY, M. AND OLLERENSHAW, J. (eds), *Land Use and the European Environment*. Belhaven.
5. EUROPEAN COMMUNITY (1988) *The Future of Rural Society*. COM (88) 501.
6. BRYDEN, J. (1989) *Why are European Policy Makers Interested in Pluriactivity?* Paper presented to Agricultural Economics Society Annual Conference.
7. EUROPEAN COMMUNITY (1988) Council Regulation (EEC) No 2052/88.
8. MACGREGOR, J. (1987) *Directions for Change: land use in the 1990s*. Paper presented to NEDO Conference, Southampton.
9. MACGREGOR, J. (1989) *Agriculture: just another industry.* MAFF press release 1/89.
10. GASSON, R. (1983) *Gainful Occupations of Farm Families*. Wye College.
11. FLETCHER, J. (ed) (1987) *Deer Farming: a realistic alternative?* BDFA and Barclays Bank.
12. A classic example is the Peat Marwick McLintock publication, *The Leisure Market* (1988).

APPENDIX I

PUBLISHED SOURCES OF INFORMATION ON ALTERNATIVE ENTERPRISES

The growing interest in alternatives has created a significant amount of writing which is not always backed up by sound evidence of viable production systems and assured markets. The available literature can be divided into a number of categories. The merits of each will be briefly examined.

Trade Journals

The broad-appeal weekly farming newspapers like *Farming News* and *Farmers Weekly* have both given much attention to alternatives. Whilst *Farming News* has been perhaps strongest in its advocacy of alternative enterprises, *Farmers Weekly* must be commended for entering the PYO field, farmhouse cheese production and farm woodland management, and describing honestly the problems that they have encountered.

There is also a wide range of more specialist journals including *Arable Farming*, *Crops*, *Dairy Farming* and *Pig Farming*. Many of these have contained articles on alternatives.

Farm Development Review, published by Cambridge Publications Ltd, offers farmers business and production advice on alternatives. It endeavours to synthesise information on particular alternatives and provides contact addresses of relevant suppliers, marketing agencies and promotional organisations.

Trade journals designed for areas like the foresty industry should not be neglected. They may contain useful articles on technical change and market intelligence that could be of relevance to the farmer with alternative enterprises.

All articles should be interpreted with care. They may at times represent free advertising for the alternative enterprise under scrutiny. It may well be in the interests of those to whom the article relates to exaggerate the potential, to ensure markets for breeding stock or to enable practitioners or consultants to sell their advice.

Advisory Body Publications

The broad range of alternatives means that the literature of many advisory organisations must be scanned. This should be more impartial than some of the evidence found in more journalistic sources. Organisations like the Forestry Commission or the Tourist Boards should be approached for catalogues of advisory publications. The reader must always beware of out-of-date costings in such publications. The type of publications available will be examined in the context of the individual bodies in Appendix II.

'Scientific' Papers and Other Similar Sources

Many farmers may feel it is unnecessary to root out scientific papers relating to alternatives. However, where expert advice is not available, scientific journals or conference proceedings may be a useful source of information. Seeking out relevant references can be a time consuming business. Most College libraries contain Abstracting Journals which include titles like Rural Recreation and Tourism Abstracts. These Abstracting Journals catalogue all relevant articles, books and conference papers under headings which can allow a systematic search through the field of past publications for a particular subject. Inevitably, however, by the time a reference has been abstracted, it is to a degree out of date.

Trade Sources of Information

Suppliers to the agricultural industry, particularly of seeds, have produced guides to alternative crops. It can be difficult to establish whether these guides are for the interest of grower or seedsman or both. Because of the infancy of many alternatives, it is important not to rely on gross margins estimated in such brochures. Nonetheless, such sources provide a catalogue of crops through which the interested farmer can browse.

APPENDIX II

The interest in alternative enterprises will frequently require recourse to advice beyond that offered by ADAS. This appendix explores the nature of advice and support offered by the main agencies. It is not comprehensive but aims to represent a selection of the key agencies. A list of smaller organisations that may be of interest to some readers is given at the end.

Business in the Community (Enterprise Agencies)

227a City Road, London, EC1V 1LX
Tel. 01–253–3716

Organisational structure: Local enterprise agencies (LEAs) are local partnerships set up by companies, local authorities and others interested in local economic development. They do not cover the whole country but have increased in number from 20 in 1981 to over 300 by 1989. There is a separate Scottish limb of Business in the Community based in Edinburgh.

Topics on which advice is given: Enterprise agencies' principal strength is the seconded advisers who are able to use their expertise to counsel those contemplating business developments. The secon-

219

dees are able to offer a very wide range of advice and may offer additional services too. Those using LEAs should be aware of the fact that they do not operate rigidly and bureaucratically and that their operation is likely to reflect the personalities in post.

Is advice charged for? No.

Grants offered and relevant financial assistance: LEAs do not give grants but might assist applicants in finding them from other sources.

Relevant publications: Promotional information is available from Business in the Community indicating the types of rural enterprise that have been supported to date.

Countryside Commission; Countryside Commission for Scotland

England and Wales—John Dower House, Cheltenham, Glos.
Tel. (0242) 521381
Scotland—Battleby, Redgorton, Perth, Scotland
Tel. (0738) 27921

Organisational structure: The Countryside Commission for England and Wales operates on a regional basis with offices in Newcastle, Manchester, Leeds, Birmingham, Cambridge, Newtown (Wales), London and Bristol. The Countryside Commission for Scotland has no regional offices. There are plans to merge the functions and organisations of the Countryside Commission and the Nature Conservancy Council in Wales and Scotland.

Topics on which advice is given: Advice can be given on a wide range of topics relating to landscape conservation and informal recreation provision. There are very few advisers in each region and advisory services are often farmed out to planning departments in the counties and districts.

Is advice charged for? No.

Grants offered and other financial assistance: Discretionary grants can be offered for a wide range of activities which contribute to landscape conservation or informal recreational provision. Grants of up to 50 per cent may be given. Management agreements and Environmentally Sensitive Area payments have proved remunerative for some farmers.

Top-up payments are available to farmers participating in set-aside in East Anglia. These discretionary grants for permanent fallow set-aside only will be given for specific management practices.

Relevant publications: The Countryside Commission publish a very wide range of literature. Much of it relates to specific aspects of recreation or conservation management and is not directly relevant to private sector, land based, recreation enterprises, although the report of the Countryside Policy Review Panel in 1987 offers a Countryside Commission view of a restructured countryside and its policy needs.

Some studies, e.g. *Small Woods on Farms, Agricultural Landscapes* and *Demonstration Farms*, are much more directly relevant. There are a number of publications detailing the grants to be offered. The Countryside Commission for Scotland offer *Countryside Conservation: a guide for farmers*, which is a clearly illustrated document detailing policies and grants available.

Farming and Wildlife Trust

National Agricultural Centre, Stoneleigh, Kenilworth,
Warwickshire, CV8 2RX
Tel. (0203) 696699

Organisational structure: The Farming and Wildlife Trust is an independent group with representatives drawn from many organisations involved in the management of the countryside. Forty-five county groups now employ a full-time adviser.

Topics on which advice is given: A broad conception of farm conservation is adopted but embraces activities like game management and woodland management. Personal advice is offered and other specialist advisers can be drawn in.

Is advice charged for? No.

Grants offered and other financial assistance: None. FWT is actively seeking funds to pursue its activities but may be able to advise farmers of appropriate sources of grant aid.

Relevant publications: FWT produces a series of advisory leaflets. Those on farm woodland and game coverts are the most relevant for alternative enterprises. Many others relate to more overtly conservation issues. A farm conservation guide published with Schering UK is obtainable from the above address.

Food from Britain

301–344 Market Towers, New Covent Garden Market, London, SW8 5NQ
Tel. 01–720–2144

Organisational structure: Food from Britain was set up in 1983, funded initially by central government but increasingly by the private sector. It has struggled to survive and has recently experienced a major restructuring which has reduced the number of regional offices to five.

Topics on which advice is given: Food from Britain aims to assist increased sales from the UK food and drink industry at home and abroad. It also acquired the responsibilities for co-operatives formerly covered by the Central Council for Agricultural and Horticultural Co-operation. Advice is given on developing export markets and developing markets for speciality foods. FFB has recently been active in determining national standards for organic produce.

Is advice charged for? Some advice, which amounts to consultancy, is, whilst some is not.

Grants offered and other financial assistance: Grants are offered for co-operatives and speciality food product marketing groups.

Relevant publications: These take the form of newsheets, eg *Enterprise Farming*, and brochures detailing the full range of services and activities of Food from Britain.

Forestry Commission

231 Corstorphine Road, Edinburgh
Tel. 031–334–0303

Organisational structure: England and Scotland are each divided into three Conservancies and Wales has one Conservancy. There are a number of private woodlands officers in these Conservancies.

Topics on which advice is given: Timber production, including, at times, considerations relating to amenity, conservation etc.

Is advice charged for? No.

Grants offered and other financial assistance: There are two principal sources of grant aid for tree growing on farms. The Farm Woodland Scheme is a MAFF/DAFS/WOAD administered scheme and the Woodland Grant Scheme is the Forestry Commission's scheme. There are fixed rates of grants with higher rates for broadleaves and grants are paid in instalments to ensure effective establishment. Grants are also offered for natural regeneration of native woodlands.

Relevant publications: The Forestry Commission produces comprehensive literature on many aspects of woodland management from silvicultural aspects to taxation of woodlands. The emphasis of much of the literature is on conifers but major recent publications cover the silviculture of broadleaved woodland.

Game Conservancy

Fordingbridge, Hampshire, SP6 1EF
Tel. (0425) 52381

Organisational structure: The Game Conservancy has specialist advisers at its headquarters and regional offices based in Gloucestershire, Yorkshire and the North of Scotland.

Topics on which advice is given: Fordingbridge provides an initial clearing house for all technical enquiries. Advice is offered on technical matters relating to shooting including habitat improvement, predator control, stocking policy etc. Advice is also given on letting and purchasing sporting rights. A number of short courses on game management are run, ranging from one to seven days' duration.

Is advice charged for? Yes to members and non-members alike.

Grants offered and other financial assistance: Not applicable.

Relevant publications: A series of advisory books and booklets is available. None covers financial and marketing aspects of game management.

Highlands and Islands Development Board

Bridge House, 20 Bridge Street, Inverness, IV1 1QR
Tel. (0463) 234171

Organisational structure: The board has four 'directorates' covering Business Services and Finance, Project Development, Marketing Development and Administrative and Central Services. Within these directorates there are individuals with specialisms in natural resource developments. The board area is divided into three main sub-regions each with a number of offices.

Topics on which advice is given: The HIDB has a very wide remit and offers advice on a very wide range of subjects. It has been active in researching alternatives itself including deer farming and is involved in a variety of development programmes.

Is advice charged for? No.

Relevant publications: The HIDB produces clear leaflets outlining the support it gives. The *How We Can Help You* booklet is supplemented by a batch of leaflets on fish farming and a series of leaflets on farm diversification is currently under preparation.

Mid Wales Development/Development Board for Rural Wales

Ladywell House, Newtown, Powys, SY16 1JB
Tel. (0686) 626965

Organisational structure: The Board works within a specified area of Mid Wales as a development agency. It is based in Newtown.

Topics on which advice is given: The Board has a reputation for concentrating its efforts into larger industrial developments rather than providing pump-priming funds for local entrepreneurs. It has not had the land development responsibilities that the HIDB possesses but professes to offer help with projects creating or maintaining employment in agriculture, forestry, fishing and tourism. However, the Board is interested in converting redundant buildings for light industrial and craft uses and may offer advice as well as grants (see below) in this area. In recent years its efforts have been far less centralised on the Newtown area.

Is advice charged for? No.

Grants offered and other financial assistance: Discretionary grants of up to 35 per cent (10 per cent higher than Agricultural Improvement Scheme) on a discretionary basis for redundant building conversions excluding tourism.

Other regional policy aids may be available to value-adding enterprises. The Mid Wales Development Grant offers variable levels of assistance.

Relevant publications: Guidelines on grant aid offered are available.

MAFF

MAFF, Whitehall, London, SW1A 2HH
Tel. 01–233–8266

Topics on which advice is given: The full range of advice is available from ADAS officers on all farming matters. Increasingly their remit is seen to extend into conservation, woodland management and game and the former socio-economic advisers have been relabelled as Rural Enterprise advisers.

Is advice charged for? Yes, apart from general advice of a preliminary nature, fees are charged.

Grants offered and other financial assistance: There have been major changes in the available range of MAFF grants in the last few years. The Farm Diversification Grants pay grants on a select range of alternatives, giving higher rates of grant to farmers less than 40 years old. Feasibility and marketing grants are also available. The major grant scheme for agricultural development, the Farm and Conservation Grant Scheme, may also be relevant especially where environmental improvement is associated with tourist or recreational developments. Two other sources of income from diversification are the Farm Woodland Scheme, which pays taxable annual payments for up to 40 years for tree planting on arable land and the payments given to farmers in Environmentally Sensitive Areas. It should be noted that under the set-aside proposals, there is scope for certain alternative enterprises.

Relevant publications: The old socio-economic advisory booklets are not being reprinted but there are detailed brochures on the grant schemes currently on offer including the *Handbook on Farm and Countryside Grants* and the document on *New Opportunities in the Countryside*.

Nature Conservancy Council

Northminster House, Peterborough, PE1 1UA
Tel. (0733) 40345

Organisational structure: There are fifteen regional offices through Britain. These local offices have some discretion in grant-aiding activities within the regions. (See also Countryside Commission.)

Topics on which advice is given: Conservation in the widest sense.

Is advice charged for? No.

Grants offered and other financial assistance: All grants are discretionary. Conditions may be imposed. All grants are given under the 1981 Wildlife and Countryside Act. Grants may be given 'to any person doing anything, which in NCC's opinion, is conducive to nature conservation or fostering the understanding of nature conservation.' It must be recognised that entrepreneurial activities will not normally be grant-aided.

Management agreements on SSSIs are drawn up between the NCC and the owner. Substantial amounts of compensation may be obtained.

Relevant publications: The NCC produces a wide and varied literature, some of which relates to farm-based conservation.

Rural Development Commission

141 Castle Street, Salisbury, Wilts., SP1 3TP
Tel. (0722) 336255

Organisational structure: The RDC is the reincarnated version of CoSIRA and the Development Commission. It has county offices in most counties containing Rural Development Areas (RDAs) and some other offices in counties without RDAs.

Topics on which advice is given: Advice is given in two main areas.
  i. Business Management—including accountancy, marketing and production management
  ii. Technical Advice—including advice from a range of traditional crafts like saddlery to new industries in the countryside.

In RDAs advice can be given on tourism. A variety of training courses is offered. Clinics on business problems are held in some areas.

Is advice charged for? Initial consultancies are free but specialist advice is charged at nominal rates.

Grants offered or other financial assistance: 25 per cent grants are given on building conversions in RDAs. In addition, grants may be given on certain other items. Loans may be offered but can only be considered as top-up finance.

Relevant publications: There are few advisory publications but *Old Buildings, New Opportunities*, 2nd edition, is a useful starting point for building conversions, and a general signposting booklet entitled *Action for Rural Enterprise* is offered which is a valuable starting point to anyone journeying on the road towards diversification.

Scottish Agricultural Colleges Advisory Services

Cleeve Gardens, Oak Bank Road, Perth, PH1 1HF
Tel. (0738) 36611

Organisational structure: The Scottish Agricultural Colleges are currently being reviewed by a Committee which may well lead to restructuring of their operations.

Topics on which advice is given: General agricultural advice will be given by conventional advisers. Socio-economic advisers exist to give advice on issues like alternatives, pensions, retirement, etc. There are also specialist advisers for specific alternatives like forestry, fish farming and deer farming. In late 1989 a Rural Enterprise Service dealing with all types of rural enterprise and not just farms is to be introduced.

Is advice charged for? Yes.

Grants offered and other financial assistance: Grants are offered by the Department of Agriculture for Scotland rather than Scottish Agricultural Colleges.

Relevant publications: The Scottish Agricultural Colleges produce a Nix equivalent detailing information on a wide range of conventional and unconventional agricultural enterprises. A publications list is available which includes titles relating to alternatives.

Tourist Boards—English Tourist Board; Scottish Tourist Board; Wales Tourist Board

England—Thames Tower, Blacks Road, Hammersmith, London, W6 9EL
Tel. 01–846–9000
Scotland—23 Ravelston Terrace, Edinburgh
Tel. 031–332–2433
Wales—Brunel House, 2 Fitzalan Road, Cardiff, CF2 1UY
Tel. (0222) 499909

Organisational structure: The ETB acts as an agency for the regional tourist boards who are the points of contact for normal enquiries for grants etc. Farmers are only one small type of fish in a large tourist sea. The Scottish Tourist Board operates centrally from Edinburgh. The HIDB covers tourist projects in their region. The Wales Tourist Board has development officers in different parts of the principality.

Topics on which advice is given: The objective of the boards is to encourage the provision of tourist facilities and the improvement of standards. Staff are willing to give advice on a wide range of tourist topics. The boards operate systems of standards.

Is advice charged for? There is no consistent pattern. Some boards charge costs; others provide free advice. The area responsibilities of the various boards in England stop any shopping around.

Grants offered and other financial assistance: All the tourist boards used to offer grant aid under the Development of Tourism Act 1969. In 1989 these grants were suspended in England but are still available in Wales and Scotland. All grants are discretionary and range from 15–40 per cent. There are signs that the boards are becoming increasingly discerning in the type of projects they will grant aid.

Relevant publications: A wide range of publications is offered by the English Tourist Board. In addition the Board produces development guides, some of which are highly relevant to small tourist operators. The Wales and Scottish Boards produce their own publications.

Other Organisations

AGRICULTURAL TRAINING BOARD, Bourne House, 32–34 Beckenham Road, Beckenham, Kent.

ASSOCIATION OF AGRICULTURE, Victoria Chambers, 16/18 Strutton Ground, London, SW1P 2HP.

BRITISH ANGORA GOAT SOCIETY, The Three Counties Showground, Malvern, Worcs, WR13 6NW.

BRITISH ASSOCIATION FOR SHOOTING AND CONSERVATION, Marford Mill, Rossett, Wrexham, Clwyd, LL12 0HL.

BRITISH COMMERCIAL RABBIT ASSOCIATION, Gauntlet Chase, Sherfield English, Hants, SO51 6JT.

BRITISH DEER ASSOCIATION, Church Farm, Lower Basildon, Reading, RG8 9NH.

BRITISH DEER FARMERS' ASSOCIATION, Holly Lodge, Spencers Lane, Berkswell, Coventry, CV7 7BZ.

BRITISH GOAT SOCIETY, 34/36 Fore Street, Bovey Tracey, near Newton Abbot, Devon, TQ13 9AD.

BRITISH GOOSE PRODUCERS ASSOCIATION, High Holborn House, 52–54 High Holborn, London, WC1V 6SX.

BRITISH HORSE SOCIETY, British Equestrian Centre, Stoneleigh, Warwicks, CV8 2LR.

BRITISH MILKSHEEP ASSOCIATION, 5b St Andrew's Square, Droitwich, Worcs.

BRITISH ORGANIC FARMERS, 86 Colston Street, Bristol, BS1 5BB.

BRITISH TROUT ASSOCIATION, PO Box 189, Fulham, London, SW6 7UT.

CARAVAN CLUB, East Grinstead House, East Grinstead, West Sussex, RH19 1UA.

COUNTRY LANDOWNERS' ASSOCIATION, 16 Belgrave Square, London, SW1X 8PQ.

DELICATESSEN AND FINE FOODS ASSOCIATION, 6 The Broadway, Thatcham, Berks, RG13 4JA.

ENGLISH COUNTRY CHEESE COUNCIL, National Dairy Centre, 5–7 John Princes Street, London, W1M 0AP.

ENGLISH VINEYARDS ASSOCIATION, 38 West Park, London, SE9 4RH.

FARM HOLIDAY BUREAU, NAC, Stoneleigh, Warwicks, CV8 2LZ.

FARM SHOP AND PICK YOUR OWN ASSOCIATION, Agriculture House, Knightsbridge, London, SW1X 7NJ.

FREE RANGE EGG ASSOCIATION, 37 Tanza Road, London, NW3 2UA.

GOAT PRODUCERS ASSOCIATION OF GB, NAC, Stoneleigh, Warwicks, CV8 2LZ.

ICE CREAM FEDERATION, 6 Catherine Street, London, WC2B 5JJ.

NATIONAL AGRICULTURAL CENTRE, Stoneleigh, Kenilworth, Coventry, CV8 2LZ.

NATIONAL ASSOCIATION OF CIDER MAKERS, Georgian House, Trinity Street, Dorchester, Dorset, DT1 1UB.

NATIONAL COUNCIL FOR VOLUNTARY ORGANISATIONS, 26 Bedford Square, London, WC1B 3HU.

NATIONAL GAME DEALERS' ASSOCIATION, 1 Belgrove, Tunbridge Wells, Kent, TN1 1YW.

NATIONAL RURAL ENTERPRISE CENTRE; National Agricultural Centre, Stoneleigh, Warwicks, CV8 2LZ.

ORGANIC ADVISORY SERVICE, Elm Farm Research Centre, Hamstead Marshall, Newbury, Bucks, RG15 0HR.

ORGANIC FARMERS' AND GROWERS' ASSOCIATION, Abacus House, Station Approach, Needham Market, Suffolk.

ORGANIC GROWERS' ASSOCIATION, 86 Colston Street, Bristol, BS1 5BB.

RARE BREEDS SURVIVAL TRUST, Fourth Street, NAC, Stoneleigh, Warwicks, CV8 2LG.

ROYAL FORESTRY SOCIETY OF ENGLAND, WALES AND NORTHERN IRELAND, 102 High Street, Tring, Hertfordshire, HP23 4AH.

ROYAL HIGHLAND AND AGRICULTURAL SOCIETY OF SCOTLAND, Edinburgh Exhibition and Trade Centre, Ingliston, Edinburgh, EH28 8NF.

SAND AND GRAVEL ASSOCIATION, 1 Bramber Court, 2 Bramber Road, London, W14 9PB.

SMALLFARMERS' ASSOCIATION, Buriton House, Newport, Saffron Walden, Essex, CB11 3PL.

SNAIL CENTRE, Plas Newydd, 90 Dinerth Road, Colwyn Bay, Clwyd, LL28 4YH.

SOIL ASSOCIATION, 86 Colston Street, Bristol, BS1 5BB.

WOODLAND TRUST, Autumn Park, Dysart Road, Grantham, Lincs, NG31 6LL.

INDEX

FARMING PRESS BOOKS

Below is a sample of the wide range of agricultural and veterinary books published by Farming Press. For more information or for a free illustrated book list please contact:

Farming Press Books, 4 Friars Courtyard
30–32 Princes Street, Ipswich 1P1 1RJ, United Kingdom
Telephone (0473) 43011

Cereal Husbandry
E. John Wibberley

A wide-ranging exposition of the principles of temperate cereal production.

Housing the Pig
Gerry Brent

Provides full guidelines enabling the pig farmer to assess proposals for new investment in buildings or equipment. Fifty detailed layouts are appraised.

Candidly Yours . . .
John Cherrington, edited by Dan Cherrington

A collection covering the years 1948 to 1988 which vividly illustrates the agricultural realities of the period.

Forage Conservation and Feeding –
4th edition
F. Raymond, and R. W. Waltham

Brings together the latest information on crop conservation, haymaking, silage making, mowing and field treatments, grass drying and forage feeding.

Pearls in the Landscape – the conservation and management of ponds
Chris Probert

A comprehensive, practical guide to the subject, including plans for creating ponds.

Farm Building Construction
Maurice Barnes and Clive Mander

A professional guide to the basic requirements in planning and constructing new farm buildings and modifying old ones.

Profitable Beef Production –
4th edition
M. McG. Cooper and M. B. Willis

Provides a concise account of the basic principles of reproduction, growth and development, nutrition and breeding. Emphasises production systems for dairy-bred beef.

Farm Machinery – 3rd edition
Brian Bell

Gives a comprehensive introduction to farm equipment, including a wide range of tractors. Over 150 plates.

Farming Press also publish four monthly magazines: *Dairy Farmer, Pig Farming, Arable Farming* and *Livestock Farming*. For a specimen copy of any of these magazines, please contact Farming Press at the address above.